Valorization of Material Wastes for Environmental, Energetic and Biomedical Applications

Valorization of Material Wastes for Environmental, Energetic and Biomedical Applications

Editor

Antonio Gil Bravo

MDPI • Basel • Beijing • Wuhan • Barcelona • Belgrade • Manchester • Tokyo • Cluj • Tianjin

Editor
Antonio Gil Bravo
Public University of Navarra
Spain

Editorial Office
MDPI
St. Alban-Anlage 66
4052 Basel, Switzerland

This is a reprint of articles from the Special Issue published online in the open access journal *Eng-Advances in Engineering* (ISSN 2673-4117) (available at: https://www.mdpi.com/journal/eng/special_issues/material_wastes).

For citation purposes, cite each article independently as indicated on the article page online and as indicated below:

LastName, A.A.; LastName, B.B.; LastName, C.C. Article Title. *Journal Name* **Year**, *Volume Number*, Page Range.

ISBN 978-3-0365-5691-8 (Hbk)
ISBN 978-3-0365-5692-5 (PDF)

© 2022 by the authors. Articles in this book are Open Access and distributed under the Creative Commons Attribution (CC BY) license, which allows users to download, copy and build upon published articles, as long as the author and publisher are properly credited, which ensures maximum dissemination and a wider impact of our publications.

The book as a whole is distributed by MDPI under the terms and conditions of the Creative Commons license CC BY-NC-ND.

Contents

Antonio Gil
Special Issue: Valorization of Material Waste for Environmental, Energetic, and Biomedical Applications
Reprinted from: *Eng* **2022**, *3*, 24–26, doi:10.3390/eng3010003 1

Ioannis Voultsos, Dimitrios Katsourinis, Dimitrios Giannopoulos and Maria Founti
Integrating LCA with Process Modeling for the Energetic and Environmental Assessment of a CHP Biomass Gasification Plant: A Case Study in Thessaly, Greece
Reprinted from: *Eng* **2020**, *1*, 2–30, doi:10.3390/eng1010002 5

Ilenia Rossetti, Francesco Conte and Gianguido Ramis
Kinetic Modelling of Biodegradability Data of Commercial Polymers Obtained under Aerobic Composting Conditions
Reprinted from: *Eng* **2021**, *2*, 54–68, doi:10.3390/eng2010005 35

Vincent H. Y. Tam, Felicia Tan and Chris Savvides
A Critical Review of the Equivalent Stoichiometric Cloud Model Q9 in Gas Explosion Modelling
Reprinted from: *Eng* **2021**, *2*, 156–180, doi:10.3390/eng2020011 51

Volker Bächle, Patrick Morsch, Marco Gleiß and Hermann Nirschl
Influence of the Precoat Layer on the Filtration Properties and Regeneration Quality of Backwashing Filters
Reprinted from: *Eng* **2021**, *2*, 181–196, doi:10.3390/eng2020012 77

Felicia Tan, Vincent H. Y. Tam and Chris Savvides
Elevated LNG Vapour Dispersion—Effects of Topography, Obstruction and Phase Change
Reprinted from: *Eng* **2021**, *2*, 249–266, doi:10.3390/eng2020016 93

Magaly Beltran-Siñani and Antonio Gil
Accounting Greenhouse Gas Emissions from Municipal Solid Waste Treatment by Composting: A Case of Study Bolivia
Reprinted from: *Eng* **2021**, *2*, 267–277, doi:10.3390/eng2030017 111

Andrés Pérez-González, Verónica Pinos-Vélez, Isabel Cipriani-Avila, Mariana Capparelli, Eliza Jara-Negrete, Andrés Alvarado, Juan Fernando Cisneros and Piercosimo Tripaldi
Adsorption of Estradiol by Natural Clays and *Daphnia magna* as Biological Filter in an Aqueous Mixture with Emerging Contaminants
Reprinted from: *Eng* **2021**, *2*, 312–324, doi:10.3390/eng2030020 123

Alex L. Riley, Christopher P. Porter and Mark D. Ogden
Selective Recovery of Copper from a Synthetic Metalliferous Waste Stream Using the Thiourea-Functionalized Ion Exchange Resin Puromet MTS9140
Reprinted from: *Eng* **2021**, *2*, 512–530, doi:10.3390/eng2040033 137

Editorial

Special Issue: Valorization of Material Waste for Environmental, Energetic, and Biomedical Applications

Antonio Gil

INAMAT2 Science Department, Campus of Arrosadia, Public University of Navarra, Building Los Acebos, 31006 Pamplona, Spain; andoni@unavarra.es

Citation: Gil, A. Special Issue: Valorization of Material Waste for Environmental, Energetic, and Biomedical Applications. *Eng* 2022, 3, 24–26. https://doi.org/10.3390/eng3010003

Received: 24 November 2021
Accepted: 22 December 2021
Published: 29 December 2021

Publisher's Note: MDPI stays neutral with regard to jurisdictional claims in published maps and institutional affiliations.

Copyright: © 2021 by the author. Licensee MDPI, Basel, Switzerland. This article is an open access article distributed under the terms and conditions of the Creative Commons Attribution (CC BY) license (https://creativecommons.org/licenses/by/4.0/).

Preface

Waste management and its recovery to provide it with added value are increasingly important lines of research that fall within the concept of a *Circular Economy*. Reducing the amount of waste that is generated is no longer only the objective; at present and in the future it is necessary to achieve a utility for this waste that contributes effectively to the scarcity of raw materials, as well as the impossibility of storing any type of waste.

The development of materials and products from industrial waste has attracted the attention of the research community for years. The physicochemical characteristics have specific impacts on the material properties and the materials and products are applied in environmental, energetic, and biomedical areas such as pollutant removal, CO_2 capture, energy storage, catalytic oxidation and reduction processes, conversion of biomass to biofuels, and drug delivery. Examples of materials are activated carbons, clays, and zeolites, among others. The aim of this Special Issue is to compile the recent advances and progress in relation to valorized materials from industrial waste and their applications in environmental, energetic, and biomedical areas.

This book contains up to eight papers published by several authors interested in the valorization of materials and in the concept of a *Circular Economy*. Voultsos et al. [1] evaluate the energetic and environmental performance of a cogeneration biomass gasification plant, situated in Thessaly, Greece, via a methodology combining process simulation and Life Cycle Assessment (LCA). Initially in the work, the gasification process of the most common agricultural residues found in the Thessaly region is simulated to establish the effect of technical parameters such as gasification temperature, equivalence ratio, and raw biomass moisture content. The gasifier model is up-scaled by the authors, achieving the operation of a 1 MW_{el} and 2.25 MW_{th} cogeneration plant. The LCA of the operation of the cogeneration unit is conducted using the performance data from the process simulation as input. Global Warming Potential and the Cumulative Demand of Non-Renewable Fossil Energy results suggest that the component which had the major share in both impact categories is the self-consumption of electricity of the plant. The results obtained by the authors suggest that plant operation in all examined conditions leads to GHG mitigation and non-renewable energy savings of approximately 0.6 kg CO_2eq/kWh_{el} and 10 MJ/kWh_{el}, respectively.

The methods to treat kinetic data for the biodegradation of various plastic materials are comparatively discussed in the work by Rossetti et al. [2]. Several samples of commercial formulates were tested for aerobic biodegradation in compost, following the standard ISO14855. Starting from the raw data, the conversion vs. time entries were elaborated using relatively simple kinetic models, such as integrated kinetic equations of zero, first, and second order; the Wilkinson model; or the Michaelis Menten approach.

Tam et al. [3] report that Q9 is widely used in industries handling flammable fluids and is central to explosion risk assessment (ERA). Q9 transforms complex flammable clouds from pressurized releases to simple cuboids with uniform stoichiometric concentration, drastically reducing the time and resources needed by ERA. Q9 is commonly believed in the industry to be conservative, but two studies on Q9 gave conflicting conclusions. This efficacy issue is important, as impacts of Q9 have real life consequences, such as inadequate engineering design and risk management, risk underestimation, etc. The authors review published data and describe additional assessment on Q9 using the largescale experimental dataset from Blast and Fire for Topside Structure joint industry (BFTSS) Phase 3B project, which was designed to address this type of scenario. The presented results show that Q9 systematically under-predicts this dataset. Finally, the authors make several observations and recommendations.

Bächle et al. [4] also report that for solid–liquid separation, filter meshes are still used across large areas today, as they offer a cost-effective alternative, for example, compared to membranes. However, particle interaction leads to a continuous blocking of the pores, which lowers the flow rate of the mesh and reduces its lifetime. This can be remedied by filter aids. In precoat filtration, these provide an already fully formed filter cake on the fabric, which acts as a surface and depth filter. This prevents interaction of the particles to be separated with the mesh and thus increases the service life of the mesh. In the work, the effect of a precoat layer with fiber lengths of cellulose on the filtration behavior is investigated.

In the following, the authors indicate that the dispersion of vapor of liquefied natural gas (LNG) is generally assumed to be from a liquid spill on the ground in hazard and risk analysis [5]. The authors also explain that the cold vapor could be discharged at a certain height through cold venting. While there is similarity to the situation where a heavier-than-air gas (e.g., CO_2) is discharged through tall vent stacks, LNG vapor is cold and induces phase change of ambient moisture, leading to changes in the thermodynamics as the vapor disperses.

Beltran-Siñani and Gil [6] present that waste generation is one of the multiple factors affecting the environment and human health that increases directly with growing population and social and economic development. The municipal solid waste disposal sites and their management create climate challenges worldwide, with one of the main problems being high biowaste content, which has direct repercussions for greenhouse gas (GHG) emissions. In the case of Bolivia, as in most developing countries, dumps are the main disposal sites for solid waste. These places are usually non-engineered and poorly implemented due to social, technical, institutional, and financial limitations. Composting plants for the treatment of biowaste appear to be an alternative solution to the problem. In this way, municipalities have implemented pilot projects with successful social results; however, access to economic and financial resources for this alternative are limited. The authors compile and summarize the Intergovernmental Panel on Climate Change (IPCC) guidelines methodology and some experimental procedures for the accounting of greenhouse gase emissions during the biowaste composting process as an alternative to its deposition in a dump or landfill.

Conventional water treatment technologies are not capable of removing estradiol from water, as indicated Pérez-González et al. [7]. Their study aims to assess a method that combines physicochemical and biological strategies to remove estradiol even when there are other compounds present in the water matrix. Na-montmorillonite, Ca-montmorillonite, and zeolite were used to remove estradiol in a medium with sulfamethoxazole, triclosan, and nicotine using a Plackett–Burman experimental design. Each treatment was followed by biological filtration with Daphnia magna. The most significant factors for estradiol adsorption were the presence of nicotine and triclosan, which favored the adsorption; the use of Ca-montmorillonite, Zeolite, and time did not favor the adsorption of estradiol.

In the final manuscript, Riley et al. [8] present a work where the extraction of Cu from mixed-metal acidic solutions by the thiourea-functionalized resin Puromet MTS9140 was studied. Despite being originally manufactured for precious metal recovery, a high selectivity towards Cu was observed over other first-row transition metals (>90% removal), highlighting the potential for Puromet MTS9140 in base metal recovery circuits. Resin behavior was characterized in batch mode under a range of pH and sulphate concentrations and as a function of flow rate in a fixed-bed setup. In each instance, a high selectivity and capacity towards Cu was observed and was unaffected by changes in solution chemistry. The work is the first detailed study of a thiourea-functionalized resin being used to selectively target Cu from a complex multi-metal solution.

Conflicts of Interest: The authors declare no conflict of interest.

References

1. Voultsos, I.; Katsourinis, D.; Giannopoulos, D.; Founti, M. Integrating LCA with Process Modeling for the Energetic and Environmental Assessment of a CHP Biomass Gasification Plant: A Case Study in Thessaly, Greece. *Eng* **2020**, *1*, 2–30. [CrossRef]
2. Rossetti, I.; Conte, F.; Ramis, G. Kinetic Modelling of Biodegradability Data of Commercial Polymers Obtained under Aerobic Composting Conditions. *Eng* **2021**, *2*, 54–58. [CrossRef]
3. Tam, V.H.Y.; Tan, F.; Savvides, C. A Critical Review of the Equivalent Stoichiometric Cloud Model Q9 in Gas Explosion Modelling. *Eng* **2021**, *2*, 156–180. [CrossRef]
4. Bachle, V.; Morsch, P.; Glei, M.; Nirschl, H. Influence of the Precoat Layer on the Filtration Properties and Regeneration Quality of Backwashing Filters. *Eng* **2021**, *2*, 181–196. [CrossRef]
5. Tan, F.; Tam, V.H.Y.; Savvides, C. Elevated LNG Vapour Dispersion—Effects of Topography, Obstruction and Phase Change. *Eng* **2021**, *2*, 249–256. [CrossRef]
6. Beltran-Siñani, M.; Gil, A. Accounting Greenhouse Gas Emissions from Municipal Solid Waste Treatment by Composting: A Case of Study Bolivia. *Eng* **2021**, *2*, 267–277. [CrossRef]
7. Pérez-González, A.; Pinos-Vélez, V.; Cipriani-Avila, I.; Capparelli, M.; Jara-Negrete, E.; Alvarado, A.; Cisneros, J.F.; Tripaldi, P. Adsorption of Estradiol by Natural Clays and Daphnia magna as Biological Filter in an Aqueous Mixture with Emerging Contaminants. *Eng* **2021**, *2*, 312–324. [CrossRef]
8. Riley, A.L.; Porter, C.P.; Ogden, M.D. Selective Recovery of Copper from a Synthetic Metalliferous Waste Stream Using the Thiourea-Functionalized Ion Exchange Resin Puromet MTS9140. *Eng* **2021**, *2*, 512–530. [CrossRef]

Short Biography of Author

Antonio Gil (Full Professor of Chemical Engineering, Universidad Pública de Navarra, Spain): Professor Gil earned his BS and MS in Chemistry at University of Basque Country (San Sebastián), and his PhD in Chemical Engineering at University of Basque Country (San Sebastián). He did postdoctoral research at the Université catholique de Louvain (Belgium) working on Spillover and Mobility of Species on Catalyst Surfaces.

The research interests of Professor Gil can be summarized as: Evaluation of the porous and surface properties of solids. Pillared clays. Gas adsorption. Energy and CO_2 storage. Pollutants adsorption. Environmental technologies. Environmental management. Preparation, characterization and catalytic performance of metal supported nanocatalysts. Industrial waste valorization.

Article

Integrating LCA with Process Modeling for the Energetic and Environmental Assessment of a CHP Biomass Gasification Plant: A Case Study in Thessaly, Greece

Ioannis Voultsos, Dimitrios Katsourinis *, Dimitrios Giannopoulos and Maria Founti

Lab of Heterogeneous Mixtures and Combustion Systems, School of Mechanical Engineering, National Technical University of Athens, Heroon Polytechniou 9, 15780 Zografou, Greece; jvoultsos@gmail.com (I.V.); digiann@central.ntua.gr (D.G.); mfou@central.ntua.gr (M.F.)
* Correspondence: dimkats@central.ntua.gr

Received: 5 August 2020; Accepted: 17 September 2020; Published: 21 September 2020

Abstract: The energetic and environmental performance of a cogeneration biomass gasification plant, situated in Thessaly, Greece is evaluated via a methodology combining process simulation and Life Cycle Assessment (LCA). Initially, the gasification process of the most common agricultural residues found in the Thessaly region is simulated to establish the effect of technical parameters such as gasification temperature, equivalence ratio and raw biomass moisture content. It is shown that a maximum gasification efficiency of approximately 70% can be reached for all feedstock types. Lower efficiency values are associated with increased raw biomass moisture content. Next, the gasifier model is up-scaled, achieving the operation of a 1 MW$_{el}$ and 2.25 MW$_{th}$ cogeneration plant. The Life Cycle Assessment of the operation of the cogeneration unit is conducted using as input the performance data from the process simulation. Global Warming Potential and the Cumulative Demand of Non-Renewable Fossil Energy results suggest that the component which had the major share in both impact categories is the self-consumption of electricity of the plant. Finally, the key conclusion of the present study is the quantification of carbon dioxide mitigation and non-renewable energy savings by comparing the biomass cogeneration unit operation with conventional reference cases.

Keywords: biomass gasification; agricultural residues; cogeneration plant; life cycle assessment; environmental impact; greenhouse gas

1. Introduction

The key priority of the European Union (EU) Climate Policy is to prevent climate change by substantially reducing greenhouse gas emissions, while encouraging other nations to contribute to this goal. The main target set by the EU for 2050 is to achieve carbon neutrality by a 60% emission reduction, realized through a 53% and 24% share of renewable energy sources and hydrogen to the final electricity demand, respectively, as well as an energy import reduction from 55% to 20% [1–3]. Solid biomass is abundant in the EU countries and accounts for more than 60% of the current renewable energy production in the EU-28 [4]. The allocation of biomass share in final energy consumption is dominated by heat (75%), followed by electricity (13%) and transport fuels (12%). CO_2 emissions from biomass usage are considered carbon neutral because the CO_2 emitted is thought to have been absorbed by the plant via a continuous natural balance of CO_2 during its life cycle (CO_2 intake for biomass growth and CO_2 emission during decay).

Although there are several types of biomass (forest residues, energy crops, etc.), this paper focuses only on agricultural residues, because the Thessaly region is a vital agricultural area of

Greece, producing several waste byproducts, which originate from various agricultural practices. This unexploited renewable energy potential can contribute to Greece's efforts in cutting down in-house CO_2 emissions. Additionally, the ash produced by the biomass usage contains high amounts of inorganic material like calcium and nitrogen compounds, and can therefore be used as a fertilizer [5–7].

The thermochemical conversion of biomass to energy can be done with direct combustion, pyrolysis and gasification [8]. Biomass gasification is a process in which solid biomass is transformed into a gaseous product via a complex series of chemical reactions and mass and energy balances. This transformation requires a gasification medium (air, oxygen or steam) by which, the hydrogen-to-carbon ratio of the final product can be higher than that of the original biomass source. Gasification is extensively used because its gaseous product, the synthesis gas (or syngas) can be burned at higher temperatures, raising the overall efficiency of the energy conversion process [8–10].

The gasification of solid fuels is not a modern invention. It was introduced for street lighting gas supply in industrialized countries in the early 19th century, as well as for liquid fuel production during World War II [11,12]. However, due to the environmental problems caused by fossil fuel consumption, biomass gasification has emerged as an energy production alternative, leading to local energy self-sufficiency and offering communities various economic and environmental benefits [8].

Experimental data about biomass gasification is readily available in the literature [13,14]. On the other hand, the complex and multi-parametric gasification process has been approached through extensive mathematical modeling efforts. Non-experimental sensitivity analysis under various operating conditions is quite commonly applied, providing a deep insight into the process at minimum cost, while also facilitating design optimization. Detailed Computational Fluid Dynamics (CFD) calculations have been thoroughly used for the modeling of biomass gasification reactors [15,16]. Given that the goal of the model is to predict the overall process efficiency and the general behavior of the gasifier operation, simple 0-D global models are preferred. Those may include reaction kinetics or adopt a thermodynamic equilibrium between the produced gases and solids [17].

The Aspen Plus process simulator is a simple and commonly used software in which 0-D models can be simulated. Lan et al. (2018) developed an integrated system model for a biomass gasification-gas turbine operation for power generation using Aspen Plus [18]. Han et al. (2017) modeled the operation of a fixed bed, downdraft gasifier, which used hardwood chips [19]. The model was validated against the experimental results from Wei et al. (2009) [20], and a sensitivity analysis was performed in order to investigate the effects of equivalence ratio, gasification temperature and moisture content on the operating conditions. Damartzis et al. (2012) modeled the operation of a bubbling fluidized bed biomass gasification unit coupled with an internal combustion engine in Aspen Plus, by using reaction kinetics instead of equilibrium models [17]. The model was validated with data from previous studies and was used to perform a sensitivity analysis to predict the system's behavior under variable gasification temperature and equivalence ratio. Marcantonio et al. (2020) [21] developed a quasi-homogeneous model in Aspen Plus to simulate biomass gasification in a fluidized–bed reactor. The model was validated for hazelnut shell gasification with various oxidizing agents, and predicted syngas compositions showed a good agreement with the experimental data.

When approaching the problem of assessing the environmental profile of energy conversion systems, it becomes evident that indirect (off-site) emissions (caused by the generation of electric consumptions or alongside the supply chain of fuels and raw materials) should be incorporated in the analysis. Therefore, a broader energetic and environmental evaluation of the gasification and energy production processes should be sought [22]. Life Cycle Assessment (LCA) is widely used to assess the environmental aspects and potential impacts of a process by using an inventory of system inputs and outputs and by interpreting the results of the inventory analysis according to the objectives of the study [23,24].

Studies on the Life Cycle Assessment (LCA) of energy production by biomass gasification and other biomass utilization techniques are widespread in the literature. Adams and McManus (2014) assessed the net energy production and the potential environmental effects of wood waste gasification in a 230 kW$_{el}$ and 500 kW$_{th}$ Combined Heat and Power (CHP) plant powered by an entrained flow gasifier, using SimaPro 7.3 [25]. Kimming et al. (2011) conducted the LCA for a 100 kW$_{el}$ CHP plant, situated in the village of Vastra Gotaland, in Sweden, in which a downdraft gasifier, fueled with willow chips, supplied synthetic gas into an internal combustion engine [26]. Yang et al. (2018) studied the Green House Gas (GHG) emissions by the major operational components of the pioneer Jiangsu Lisen 20 MW CHP plant, which is powered by four fluidized bed gasifiers and four exhaust heat recovery boilers and is situated in the city of Yancheng, China [27]. Tagliaferri et al. (2018) carried out the LCA of a 2 MW$_{el}$ and 8 MW$_{th}$ Organic Rankine Cycle (ORC) CHP plant, supplied by forest biomass, which powers the Heathrow terminals 2 and 5, in order to assess the energy conversion process which positively contributes the most to the environmental impact of the plant [28]. Nguyen and Hermansen (2014) conducted an LCA study of all processing steps (cultivation, collection and pre-process and thermochemical conversion to electricity) of miscanthus gasification for electricity and heat production [29]. Guerra et al. (2017) identified the thermodynamic and environmental effects of scaling up existing cogeneration units in order to use sugarcane biomass as fuel via a plant LCA [30].

Most LCA studies on energy production via biomass gasification use fixed data from previous simulations and do not benefit from the advantages of detailed process modeling. As a consequence, results are not case-specific and are not adapted to the plant configuration and the feedstocks involved. Furthermore, the use of literature data does not promote comprehensiveness, since the influence of the variation of key operational parameters (equivalence ratio, gasification temperature and raw biomass moisture content) to the biomass gasification process are not considered. Overall, there are only few reports coupling process simulations together with LCA, and most of them are not directly linked to biomass gasification [22,31,32]. However, Hamedani et al. (2018) [33] performed an LCA study to evaluate the environmental profile of a real, small-scale, biomass-based hydrogen and electricity production system. They specifically focused on the effect various aspects and alternative scenarios of the gasification process have on the examined impact categories. Furthermore, Hamedani et al. (2019) [34] combined data envelopment analysis (DEA) and LCA in order to assess the sustainability of bioelectricity production by vineyard waste biomass gasification. The primary objective of this work is to introduce a comprehensive and integrated model based on the coupling of Aspen Plus and SimaPro, with the ability to assess the energetic and environmental performance of a prospective 1 MW$_{el}$ and 2.25 MW$_{th}$ cogeneration biomass gasification plant in Thessaly, Greece. The developed model provides the necessary flexibility to simulate all types of gasification layouts and operating conditions (different cases of biomass quality, equivalence ratio, gasification temperature, electric and thermal output of the cogeneration plant). The advantages of the approach are showcased via a sensitivity analysis that establishes the effect of gasification temperature, equivalence ratio and raw biomass moisture content over the gasification efficiency and the quality of the produced syngas. The obtained data are coupled with local biomass availability scenarios and are used as inputs to the Life Cycle Assessment of the cogeneration unit, so as to highlight its environmental benefits.

2. Materials and Methods

2.1. Case Study Description

The prospective CHP power plant is considered to power a village of 1500 residents located in Thessaly, Greece, which is the leading area of Greek large-scale farming and farming-related industry (fertilizers, agricultural tooling and machinery production, dairy and cereal production). Therefore, biomass in the form of agricultural residues is abundant and can potentially be used for the production of electricity and heat for local villages and industries. Given that the plant installed capacity depends on biomass availability and on costs associated with plant construction and operation, biomass collection, storage and transport, the biomass gasification the CHP plant is proposed to have an electricity output of 1 MW$_{el}$, in order to provide local self-sufficiency at a reasonable cost [35,36].

Certain assumptions had to be made regarding the specific CHP technologies considered in the integrated process and Life Cycle Assessment modeling. The power-to-heat ratio as well as the electrical and thermal efficiency of the investigated plant were taken from data available in the literature on an actual CHP plant, situated in Güssing, Austria. The Güssing plant is a 2 MW_{el} state-of-the-art and well optimized unit, operating since 2002 [37]. Its basic operational parameters (power-to-heat ratio, electrical, thermal and total efficiency) are considered in the 1 MW_{el} Greek prospective plant. The assumed operational parameters in the prospective Thessaly plant are summarized in Table 1.

Table 1. Operational parameters assumed in the prospective Thessaly plant [37].

Operational Parameter	Value
Electrical Power (MW_{el})	1
Power-to-heat ratio	2.25
Electrical Efficiency (%)	25
Thermal Efficiency (%)	56.3
Total Efficiency (%)	81.3

Wheat straw, corn stover, cotton stalk, olive branches and almond prunings are identified as the most common agricultural residues in Thessaly, Greece [38]. Their most significant characteristics are described in the works or Rentizelas et al. (2009), Voivontas et al. (2001) and Papadopoulos and Katsigiannis (2002) [38–40] and are presented in Table 2. Rentizelas et al. (2009) also suggested that, in order to power a 1 MW_{el} tri-generation plant, based in Thessaly, Greece, 52,849 m^3 of agricultural residues of all types are required per year. This amount of biomass is considered to be the total available supply to the prospective CHP biomass gasification plant examined in this study.

Table 2. Characteristics of the most common agricultural residues in Thessaly, Greece [38–40].

Characteristic	Wheat Straw	Corn Stover	Cotton Stalk	Olive Branches	Almond Prunings
Residue yield (t/ha)	2.97	7.17	5.47	2.82	6.21
Residue availability factor (%)	15	30	70	90	90
Exploitable residue (t/ha)	0.45	2.15	3.83	2.54	5.59
Moisture (%)	20	50	30	35	40
Residue density (kg/m^3)	140	200	200	250	300
Availability	July–Aug.	Nov.–Dec.	Oct.–Nov.	Nov.–Feb.	Dec.–Feb.

However, since the contribution of each biomass type to the total feedstock demand is not available, specific assumptions have been made. The mass residue yield is converted to volume yield by considering a constant density for each residue (Table 2). As a result, the individual percentage of each feedstock to the total biomass volume is calculated by dividing its volume residue yield to the total residue demand. Consequently, the emerging annual volume of each feedstock can be converted back to mass units, using the constant residue density. The results of this analysis are presented in Tables 3 and 4 and provide a feedstock availability scenario for the simulated CHP biomass gasification plant. They are used as an input in the LCA study.

Table 3. Individual contribution of common Thessaly feedstocks to the total annual volume of supplied biomass.

Biomass Type	Exploitable Residue Mass Yield (kg/ha)	Exploitable Residue Volume Yield (m³/ha)	Feedstock Contribution to Total Volume
Cotton stalk	3.83	19.15	30.9
Corn stover	2.15	10.75	17.4
Olive branches	2.54	10.16	16.4
Almond prunings	5.59	18.63	30.1
Wheat straw	0.45	3.21	5.2
Total	14.56	61.9	100

Table 4. Annual available volume and mass of common Thessaly feedstocks.

Biomass Type	Feedstock Annual Volume (m³)	Feedstock Annual Mass (kg)
Cotton stalk	16,330	3266
Corn stover	9196	1839
Olive branches	8667	2167
Almond prunings	15,908	4772
Wheat straw	2748	385
Total	52,849	12,429

As data for the Greek agricultural residues examined in this work were not available, the Phyllis2 database of the Energy Research of the Netherlands (ECN) was used, which contains information about the composition of different biomass feedstocks used for biogas, biochar and torrefied biomass production. The proximate and ultimate analyses of the five biomass types considered, as well as their Phyllis2 database IDs, are presented in Tables 5 and 6 [41].

Table 5. Proximate analysis (% wt, dry basis) of common Thessaly feedstocks.

Proximate Analysis	Wheat Straw	Corn Stover	Cotton Stalk	Olive Branches	Almond Prunings
Moisture Content	9.19	5	7.37	13.83	11.40
Volatile Matter	75.54	78.1	75.69	81.37	79.01
Fixed Carbon	16.22	14.55	19.26	16.4	19.11
Ash	8.24	7.35	5.05	2.23	1.88
LHV (MJ/kg, dry)	16.44	17.73	15.96	17.63	19.47
Phyllis2 ID	703	889	None/[42]	3347	3343

Table 6. Ultimate analysis (% wt, dry basis) of common Thessaly feedstocks.

Ultimate Analysis	Wheat Straw	Corn Stover	Cotton Stalk	Olive Branches	Almond Prunings
C	45.02	46.5	46.42	47.68	49.17
H	5.66	5.81	4.95	5.85	5.92
N	0.91	0.56	1.13	0.58	0.62
O	39.72	39.67	42.45	43.56	42.41
Ash	8.24	7.35	5.05	2.23	1.88

2.2. Gasification Modelling

Process modelling provides an essential tool for the simulation of the biomass gasification unit powering the CHP plant. By developing simple, yet accurate mathematical modeling tools, further design and optimization studies can be achieved [17]. In the present study, a computational model has been developed in Aspen Plus in order to simulate a standard case of a fixed bed downdraft gasifier and to accurately predict the produced syngas composition as well as the overall gasification process efficiency at various operating conditions. This type of gasifier is considered to be more suitable for small-scale (up to 10 MW$_{th}$) and decentralized applications. Furthermore, due to the high temperatures identified in the oxidation zone, tar cracking reactions are promoted and the produced syngas has a low tar content [43–45].

The gasification process takes place in four stages; (a) Drying (less than 150 °C), (b) Pyrolysis (150–700 °C), (c) Oxidation (700–1500 °C) and (d) Reduction (800–1100 °C) [10]. In the drying stage, raw biomass is stripped out of a high portion of its moisture content, which is transformed into steam [10]. In the pyrolysis stage, the volatile content of biomass is vaporized into a mixture of various substances like H_2, CO, CO_2 and CH_4. Furthermore, high molecular mass hydrocarbons are produced. They are considered as tars and char, a solid residue, which is considered mainly as carbon [13,14]. In the oxidation stage, oxygen of the gasification medium reacts with the combustible products of pyrolysis, resulting in the formation of CO_2 and H_2O [10]. In the reduction stage, which is mainly endothermic, gaseous products react via a series of reactions like (i) the Water-Gas Shift Reaction, (ii) the Boudouard Reaction and (iii) methanation [15]. Also, they come into contact with the solid char, and thus a series of solid-gas reactions occur. Due to the endothermic nature of the reduction stage, its temperature is significantly lowered. The final product of the gasification process is the synthesis gas (or syngas), which is a mixture of CO, CO_2, H_2 and CH_4 [14]. The main reactions taking place in the gasifier are presented in Table 7. The reaction enthalpies for the single and multi-phase reactions are taken from the literature [10].

Table 7. Single-phase and multi-phase gasification reactions [10].

Reaction	ΔH (kJ/mol)	Reaction Number	Reaction Name
Oxidation Stage			
$C_{(s)} + O_2 \leftrightarrow CO_2$	−393	R-1	Char Oxidation
$C_{(s)} + \frac{1}{2} O_2 \leftrightarrow CO$	−112	R-2	Char Partial Oxidation
$CO + \frac{1}{2} O_2 \leftrightarrow CO_2$	−283	R-3	CO Oxidation
$H_2 + \frac{1}{2} O_2 \leftrightarrow H_2O$	−242	R-4	H_2 Oxidation
Reduction Stage			
$CO + H_2O \leftrightarrow CO_2 + H_2$	−41	R-5	Water Gas Shift
$CH_4 + H_2O \leftrightarrow CO + 3 H_2$	+206	R-6	Methane Steam Reforming
$C_{(s)} + CO_2 \leftrightarrow 2 CO$	+173	R-7	Boudouard
$C_{(s)} + 2 H_2 \leftrightarrow CH_4$	−75	R-8	Methanation
$C_{(s)} + H_2O \leftrightarrow CO + H_2$	+131	R-9	Char Water Gas

The simplification of the process modeling and conformity with Aspen Plus simulation software required several assumptions. First of all, due to the 0-D nature of the aforementioned software, the fluid mechanics equations that characterize the process were not taken into account, and a uniform distribution of gases was considered inside the gasifier. The target was to provide a simple biomass gasification model with the ability to evaluate the influence of the basic parameters affecting the process, without taking into account the gasifier's configuration and dimensions. Towards this aim, an equilibrium approach has been implemented. The process was examined considering steady-state and isothermal conditions, and the gasification medium was air at 1 atm and 25 °C. The products of biomass devolatilization were H_2, CO, CO_2, CH_4 and H_2O. Tars produced during the pyrolysis stage were not taken into account. This is considered to be a reasonable assumption for downdraft gasification [46]. Furthermore, sulfur and nitrogen reactions are not considered in this study [46]. Char is modeled as the sum of fixed carbon and ash which are both reported in the respective biomass proximate analyses [46]. Thus, the percentage of carbon in volatiles was calculated by subtracting the fixed carbon percentage from the percentage of the total carbon in the biomass (included in the ultimate analysis of the dried biomass) [17].

In order to assess the energetic performance of the considered gasification process, the influence of the air equivalence ratio on the quality of the produced syngas must be evaluated. Furthermore, the lower heating value of the produced syngas and the efficiency of the gasification process must be determined. It should be noted that the air equivalence ratio plays a major role in the quantity and the quality of the produced synthesis gas. Previous studies concluded that an optimum operation and quality of produced syngas can be expected for an equivalence ratio ranging between 0.2 and 0.4 [12,47]. Equivalence ratios lower than 0.2 promote pyrolysis conditions, whereas values higher than 0.4 promote oxidation conditions. The equivalence ratio is calculated by the following equation:

$$ER = (air\ feed\ (kg)/biomass\ feed\ (kg))/(A/F)_{stoic}, \qquad (1)$$

where $(A/F)_{stoic}$ is the stoichiometric air to biomass ratio, which is calculated by Equation (2) (γ_i are the mass fractions of C, H, S and O elements in the dried biomass) [48]:

$$(A/F)_{stoic} = 11.48\gamma_C + 34.194\gamma_H + 4.3\gamma_S - 4.308\gamma_O. \qquad (2)$$

The lower heating value of the synthesis gas is calculated for standard conditions (0 °C, 1 atm) using Equation (3) [49]:

$$LHV_{syngas} = (30X_{CO} + 25.7X_{H2} + 85.4X_{CH4}) \cdot 4.2/1000\ (MJ/Nm^3) \qquad (3)$$

where X_i are the volume fractions of CO, H_2 and CH_4 in the synthesis gas. Then, it is converted to actual gasification conditions (25 °C, 1 atm). Evidently, the gasification's main target is to produce syngas with high LHV. Thus, increased values of the aforementioned species volume fractions are anticipated.

The gasification efficiency corresponds to the chemical efficiency: the greater part, that takes into account only the chemical energy and the enthalpy that is associated with the thermal energy. Chemical efficiency is named cold gas efficiency and represents the chemical energy content of the synthesis gas. It is calculated via Equation (4):

$$n_{CG} = [V_{syngas}\ (m^3 \cdot h^{-1}) \cdot LHV_{syngas}\ (MJ \cdot m^{-3})]/[m_{biomass}\ (kg \cdot h^{-1}) \cdot LHV_{biomass}\ (MJ \cdot kg^{-1})], \qquad (4)$$

where V_{syngas} and $m_{biomass}$ were the volume flow of the produced syngas and the biomass mass feed respectively.

The development of the gasification model is based on using different modules of the Aspen Plus software in order to simulate the gasifier operation, followed by an after-treatment of the produced gas, which consists of (a) cleaning the synthesis gas using cyclones and (b) cooling it with the use of heat exchangers. The gasifier is modeled via a combination of different blocks, each corresponding to a specific gasification step (i.e., drying, pyrolysis, oxidation, reduction). The simulation flowchart of the process is presented in Figure 1. Input data used for the simulation are summarized in Table 8. For the modeling of the behavior of gases, the Peng–Robinson equation of state was used.

The biomass drying step is simulated via the stoichiometric reactor RSTOIC. The drying process achieved a total drying of the biomass feed at a temperature of 150 °C.

After being stripped from the moisture, dried biomass enters an RYIELD block in order to be decomposed into its constituent, conventional components (carbon, hydrogen, oxygen, nitrogen, sulfur, ash). Decomposition calculations are based on the ultimate and proximate analyses of the five biomass types considering a 100% conversion. A temperature of 500 °C was selected. The RYIELD block practically corresponds to a simplified pyrolysis step.

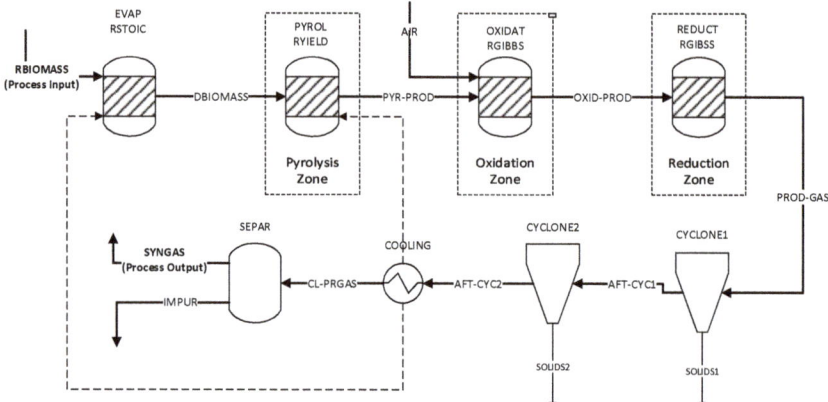

Figure 1. Biomass gasification simulation flowchart (solid lines: mass flows, dashed lines: energy flows).

Table 8. Gasification process simulation input data.

Simulation Input Data	Value
Gasification Air Pressure	1 atm
Gasification Air Temperature	25 °C
Biomass Feed Temperature	25 °C
Feed Moisture Content	As described on Table 5
Final Moisture Content	0 %
Drying Temperature	150 °C
Drying Pressure	1 atm
Pyrolysis Temperature	500 °C
Pyrolysis Pressure	1 atm
Equivalence Ratio (0.025 increment)	0.2–0.4
Oxidation Temperature (50 °C increment)	800–1200 °C
Reduction Temperature (50 °C increment)	600–1000 °C
Syngas Cooling Temperature	25 °C

Two RGIBBS reactors have been implemented for the modeling of the oxidation and reduction zones, respectively. Thus, it is assumed that, in both stages, all compounds involved have reached chemical equilibrium. The first reactor is fed with air to simulate the oxidation zone, and the second one simulates the reduction zone at a lower temperature than that of the oxidation zone. According to various literature sources, equilibrium models, implementing RGIBBS reactors, tend to overestimate CO and H_2 and underestimate CH_4 and CO_2 volume fractions [50,51]. Various solutions have been proposed. In order to tune the syngas composition to more realistic figures, Fernandez-Lopez et al. (2017) specifically defined the chemical equilibrium of reactions 5 and 6 to occur at a different temperature than the overall temperature of the reduction zone block [52]. This solution, although it produces more accurate results, has limited applicability, since it is compliant only with the experimental data used for the tuning. Atnaw et al. (2018) suggested that, in a downdraft gasifier, the chemical equilibrium of the reduction zone reactions (presented in Table 7), which are highly endothermal and therefore lower the gasification temperature, should be set at a temperature 200 °C lower than that of the oxidation zone reactions [53]. This approach is expected to produce reasonable results for fixed bed as well as for fluidized bed gasifiers under various operating conditions and for different biomass feeds. Therefore, it has been implemented in the developed model.

The next step of the gasification modelling is the refinement process of the product gas. At first, it enters two CYCLONE blocks that simulate the separation of the solid impurities (ash, occasional char residues) from the gas phase. Then, the gas stream enters a COOLING block, which simulates the operation of a heat exchanger. The synthesis gas is cooled to a final temperature of 30 °C, with respect to being potentially used as a fuel in an internal combustion engine implemented in the prospective CHP plant. The generated heat can be used for the drying process [54]. Finally, N_2, which is contained in the air feed and considered to be unreactive, is removed via the use of a SEPARATOR block, which simulates syngas cleaning via a one-step simplified process. From the above processes, the purified syngas produced is ready to power internal combustion engines or boilers in power plants.

2.3. LCA Modelling

2.3.1. LCA Framework

Life Cycle Assessment is one of the most developed and widely used methods for the quantification of the amount of materials and energy used for a process, as well as its emissions, by considering the complete supply chain of the goods and services involved [55]. It also contributes to the detection and the refinement of specific system activities, which have the most severe environmental impact. The whole LCA process follows the ISO 14040 and 14044 protocols.

2.3.2. Goal and Scope Definition–Functional Unit-System Boundaries

The aim of the study is to assess the environmental impacts associated with the operation of a combined heat and power biomass gasification power plant, powered with the most dominant agricultural residues of the Thessaly region, Greece, to identify the environmental "hotspots" of the whole energy production process and to compare the environmental burdens of syngas produced by the plant versus those of the Greek natural gas supply chain. The environmental footprint of the considered biomass gasification CHP plant was compared with conventional energy production alternatives, namely:

1. Electricity from the grid of mainland Greece according to the current energy mix.
2. Electricity from the grid of mainland Greece according to the current 2050 policy projections.
3. Electricity from a natural gas internal combustion engine on CHP mode.

The functional unit of the study was 1 kWh of electrical output. The thermal energy produced by the plant, as it is described in ISO 14040, was treated as an avoided product of a conventional condensing boiler, fired by natural gas. An additional functional unit of 1 MJ energy content was considered in the case of comparing the syngas production versus natural gas supply. The software used for the Life Cycle assessment was Simapro 7.2, which was equipped with Ecoinvent 2.0 database.

In this study, a "Cradle to Gate" Life Cycle Assessment was performed. Cotton stalk was determined to be the main biomass type used, so the plant is thought to be installed in the cotton stalk production area. It was assumed that all other biomass types were transported to the plant for an average distance of 30 km via 28 t trucks. The trucks were loaded in the biomass production site and unloaded in the CHP plant via loading machines. Cotton stalk was handled inside the plant premises via proper handling equipment. The main operations of the evaluated CHP biomass gasification plant that were considered in the LCA study are (i.) Loading of agricultural residues to trucks, (ii.) Transportation to the plant, (iii.) Unloading at the plant, (iv.) Plant construction and operation and (v.) Biomass handling. The system boundaries are presented in Figure 2.

The cultivation of biomass was not included in this study, because agricultural residues are wastes of farming activities, and thus their cultivation does not contribute to the environmental burdens of the system examined. The life span of the plant was determined to be 20 years, while plant decommissioning and ash treatment were not taken into account. The electricity required for the operation of the gasification plant is thought to be obtained directly from the mainland Greek electricity grid.

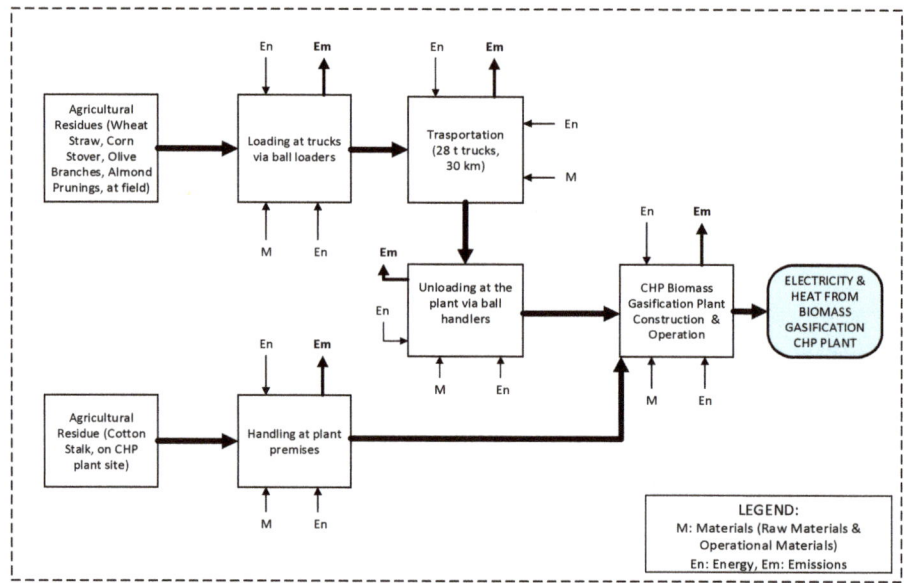

Figure 2. LCA approach: system boundaries.

2.3.3. Life Cycle Inventory Analysis

The energy production via biomass gasification includes various subprocesses which involve energy and material exchange between the CHP plant, the technosphere and the environment. As a result, mass and energy flows between all operations which characterize the CHP biomass gasification plant (shown in Figure 2) should be modeled and eventually quantified. The main inventory data used in the LCA study of this work are summarized in Table 9.

Since the aim of this study is to provide an insight into the environmental impact of the simulated CHP biomass gasification plant, life cycle inventory data were obtained from the Ecoinvent 2.0 database, which was modified in order to include case-specific data such as the performance of the simulated biomass gasification plant and all information associated with the local biomass supply chain [56].

As already described, the simulated CHP plant has an output power of 1 MW_{el} and 2.25 MW_{th} and is supplied with syngas, which is the output of the gasification process simulated in Aspen Plus V8.8. In this study, the plant was considered to run only at full load, whereas partial load scenarios were not examined. In order to couple the operation of the CHP plant with the operation of the gasifier, the major data input requirements were the characteristics (cold gas efficiency, syngas and raw biomass LHV) of the optimal operating points in terms of maximum gasification efficiency as well as the variation of the aforementioned parameters with respect to raw biomass moisture content variation. Thus, the designed load of the CHP plant (3.25 MW) determines not only the amount of the required syngas volume flow from the gasification process but also the biomass feed which is supplied into the reactor for every raw biomass moisture content scenario. For simplicity reasons, all feedstocks examined in different moisture content scenarios have the same initial moisture content, which varies from 0% (optimal operating point) to 30% (most unfavorable scenario).

Table 9. Life Cycle Inventory data for the production of electricity and heat from the CHP biomass gasification plant considered in this work for plant optimal operation (Moisture content—MC = 0%) and for the most unfavorable scenario examined (30% MC for initial feedstock).

Components	Optimal Plant Operation (MC = 0% for All Feedstocks)	Most Unfavorable Scenario (MC = 30% for All Feedstocks)
Inputs		
Materials		
Cotton stalk (kg)	0.374	0.471
Almond prunings (kg)	0.321	0.687
Olive branches (kg)	0.248	0.313
Corn stover (kg)	0.209	0.265
Wheat straw (kg)	0.0437	0.0547
Loading/Unloading diesel fuel (kg)	0.0075	0.0114
Energy		
Electricity (MJ)	0.0559	0.0559
Transport (tkm)	0.03	0.03
Syngas Energy Allocation according to feedstock		
Cotton stalk (MJ)	4.16	3.46
Almond prunings (MJ)	4.12	5.76
Olive branches (MJ)	3.12	2.59
Corn stover (MJ)	2.49	2.16
Wheat straw (MJ)	0.518	0.432
Output		
Electricity(kWh)	1	1
Heat for final use (MJ)	8.1	8.1
Emissions		
CO (g)	0.226	0.479
NO_x (g)	3.46	5.43
Avoided Products		
Electricity-Greek mixture (kWh)	1	1
Heat from natural Gas (MJ)	8.1	8.1

The equipment used for loading biomass into trucks and unloading it onto the plant site was considered to be a bale loader. The same equipment is used to handle in-house cotton stalk residues inside the CHP plant. It was modeled via "Baling" and "Loading Bales" processes, which were modified in order to calculate all flows per mass unit and to include both loading and unloading processes.

The CHP plant was assumed to be situated in the location of production of cotton stalk, because it was considered as the main biomass type in this study. All other residues were transferred with fleet average, 28 ton trucks, from distances lower than, or equal to, 30 km, in order to minimize transfer costs [38]. For simplicity, the transfer distance was considered to be constant and no intermediate distances were examined. Thus, the ton-kilometer value used in the LCA study was 0.03. Trucks return empty to the loading site, so a loading factor of 50% was considered.

The operation of the biomass gasification plant was modeled via the datasheet "Synthetic gas, from wood, at fixed bed gasifier", assuming that only the gasification plant operation contributed to the total environmental burdens of the plant. This could also be justified by the fact that all CO_2 emissions were considered to be carbon-neutral, and other byproducts and byprocesses of the CHP plant, such as the disposal of bottom ash, were not considered in this study. The datasheet was modified to account for the current and 2050 projection electricity mixture and relates the biomass quantity required in order to produce 1 m^3 of syngas. Furthermore, the volume and energy content of syngas were associated using custom datasheets via the syngas LHV, for every feedstock, moisture content and electricity grid mixture examined.

However, an assumption should be made regarding the fact that the energy from syngas produced from each feedstock was taken into account in the production of electricity and heat from the CHP biomass gasification plant. For this reason, it was assumed that the energy released by syngas burn could be allocated to the five feedstocks which were simulated in this work. Allocation to each feedstock was done by calculating the normalized working hours of the plant for each feedstock examined, using as input their annual available mass (presented in Table 4). More specifically, by calculating the required feed rate (kg/h) of each feedstock in order to reach the designed power output of the plant and by dividing it by the annual supply of its type, the plant's annual working hours for each feedstock could be determined. The normalized working hours for each feedstock could be calculated as a fraction of each feedstock working hours to the total plant working hours for the two operational scenarios examined. The calculation of syngas energy allocation to different feedstocks examined in this study is summarized in Table 10.

Table 10. Allocation of syngas energy per feedstock, according to the annual biomass availability figures and required biomass feed for the designed energy output.

Feedstock	Allocation of Syngas Energy Per Feedstock (%)	
	Optimal Plant Operation (MC = 0%)	Most Unfavorable Scenario (MC = 30% for All Feedstocks)
Cotton stalk	28.9	24
Almond prunings	28.6	40
Olive branches	21.5	18
Corn stover	17.3	15
Wheat straw	3.6	3

Finally, the energy conversion in the CHP plant was simulated in a custom datasheet, in which electricity from the CHP plant was associated with the required energy from the syngas energetic mixture via the plant's electrical efficiency (presented in Table 1). The thermal energy produced in the CHP plant was included in the aforementioned datasheet as the heat produced by a conventional condensing boiler, fired by natural gas, which is treated as an avoided product.

The electricity production from the CHP biomass gasification plant was compared with the electricity from the grid of mainland Greece. The operation of the gasification plant was assumed to require electricity from the grid. The present energy mixture of Greece, as well as the 2050 projection under current energy policies, were obtained from the DAS Monthly Reports of the Greek Operator of Electricity Market and are presented in Figure 3a,b [Source: http://www.lagie.gr]. The two electricity mixtures examined were inserted in the Greek Electricity mix datasheet.

Furthermore, for the conventional alternative of the natural gas internal combustion engine, which is used for electricity and heat production, a Deutz TBG 620K genset was considered. Technical specifications for the aforementioned engine were obtained from the official site of the manufacturer and are presented in Table 11. The engine was modeled in a custom datasheet, which connected the electrical output of the engine with the natural gas energy required via the existing "Natural Gas, burned in Cogen 1MW$_{el}$ lean burn" datasheet and the electrical efficiency of the engine. It should be mentioned that the thermal power produced by the engine is included in the study as an avoided product. It should be noted that the datasheet, which modeled the natural gas burn, included combustion, plant operation and natural gas supply chain emissions.

Finally, the use of natural gas in the grid electricity mixture and the CHP internal combustion engine required a modeling of the Greek natural gas supply chain. According to the Greek Public Gas Corporation (DEPA), high-pressure natural gas is transferred via pipelines (83%) and LNG is transferred via ships (17%) [https://www.depa.gr/natural-gas-commerce]. In this work, for the sake of simplicity, the allocation of natural gas originating from different sources into the total mixture was determined by its energy and not by its quantity, assuming that the pipeline natural gas and the LNG have the same lower heating value.

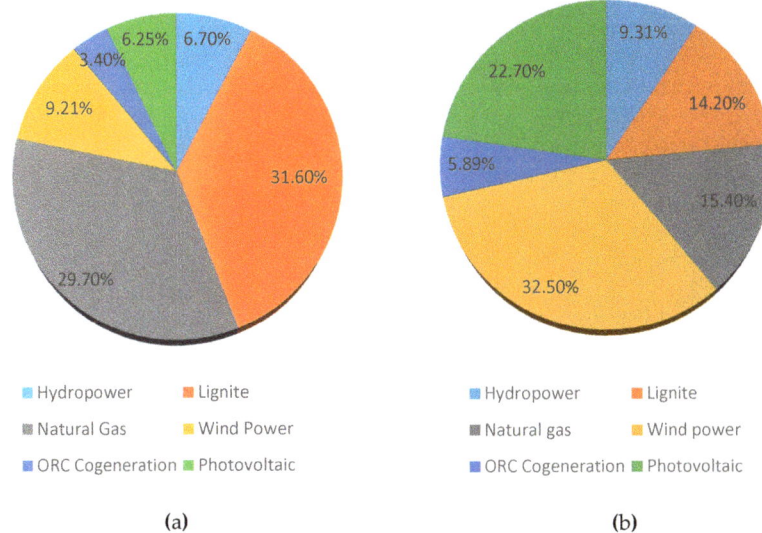

Figure 3. (**a**) Current Electricity mixture of mainland Greece [Source: DAS Monthly Reports, http://www.lagie.gr/en/market/market-analysis/das-monthly-reports/]; (**b**) Projection for the 2050 Electricity mixture of mainland Greece, under the current policy scenario [Source: http://www.lagie.gr].

Table 11. Technical specifications of Deutz TBG 620K natural gas internal combustion engine, which is simulated in this work [Source: http://www.deutz.com/].

Technical Specification	Value
Electrical Power (kW)	1022
Power to Heat Ratio	0.887
Electrical Efficiency (%)	40.2
Thermal Efficiency (%)	45.3
Total Efficiency (%)	85.5

2.3.4. Life Cycle Impact Assessment Method

The impact categories assessed in this study are:

- Global Warming Potential (GWP) (units: kg CO_{2eq}/ kWh_{el}) It was assessed via the IPCC GWP 100a method [57].
- Cumulative Energy Demand of Non-Renewable Fossil Energy (units: MJ of fossil energy/kWh). It was assessed using the Cumulative Energy Demand V1.07 method [58]

3. Results and Discussion

3.1. Gasification Modeling Results

3.1.1. Model Validation

The results of the modeled biomass gasification process modeling via Aspen Plus were used as an input for the Life Cycle Assessment of the prospective CHP plant. The influence of various operating parameters such as gasification temperature, equivalence ratio and raw biomass moisture content to the gasification cold gas efficiency was examined. At first, the performance of the developed model was assessed by comparing computational results to available experimental data. Given that the cold gas efficiency of the gasification process is defined through the syngas LHV, the developed modeling approach (as presented in the previous section) was evaluated by comparing the predicted syngas LHV against measured values from Atnaw et al. (2018) and Damartzis et al. (2012) [17,53].

Table 12 presents a summary of the layout and experimental conditions of the aforementioned studies. Proximate and ultimate analyses of the considered feedstocks are shown in Table 13. Simulations have been performed at different equivalence ratios for each case: 0.35 for Atnaw et al. and 0.2 for Damartzis et al. The comparison between the experimental syngas LHV and computational results is depicted in Table 14. As can be seen, predicted LHVs are in good agreement with the respective experimental values. Discrepancies are in the range of 10–15%, which is reasonable when equilibrium models are used for gasification modeling [21,59]. Furthermore, a comparison between syngas component yields as provided by the aforementioned studies and those produced by the developed model at a temperature of 850 °C is shown in Figures 4 and 5.

Table 12. Layout and experimental conditions of Atnaw et al. [53] and Damartzis et al. [17].

Experimental Conditions	Atnaw et al. (2018) [53]	Damartzis et al. (2012) [17]
Gasifier Type	Fixed bed Downdraft	Bubbling Fluidized Bed
Feedstock	Oil Palm Frond	Olive kernel
Thermal Power (kW_{th})	50	5
Gasification Medium	Air	Air
Equivalence Ratio	0.35	0.2
Gasification Temperature	500–1200	750–850

Table 13. Proximate and ultimate analysis of feedstocks used by Atnaw et al. and Damartzis et al. [17,53].

Proximate Analysis (% wt, Dry Basis)			Ultimate Analysis (% wt, Dry Basis)		
Literature Data Title	Atnaw et al. (2018)	Damartzis et al. (2012)		Atnaw et al. (2018)	Damartzis et al. (2012)
Moisture Content	8	4.59	C	44.58	48.59
Volatile Matter	83.5	75.56	H	4.53	5.73
Fixed Carbon	15.2	16.39	N	0.79	1.57
Ash	1.3	3.46	O	48.8	44.06
LHV (MJ/kg_{dry})	15.59	18	Ash	1.3	3.46

Table 14. Syngas lower heating value comparison between experimental data ([17,53]) and computational results.

Literature Data Title	Experimental LHV (MJ/m^3)	Predicted LHV by Model (MJ/m^3)	Difference (%)
Atnaw et al. (2018)	5	4.43	11.4
Damartzis et al. (2012)	5.14	5.91	−15

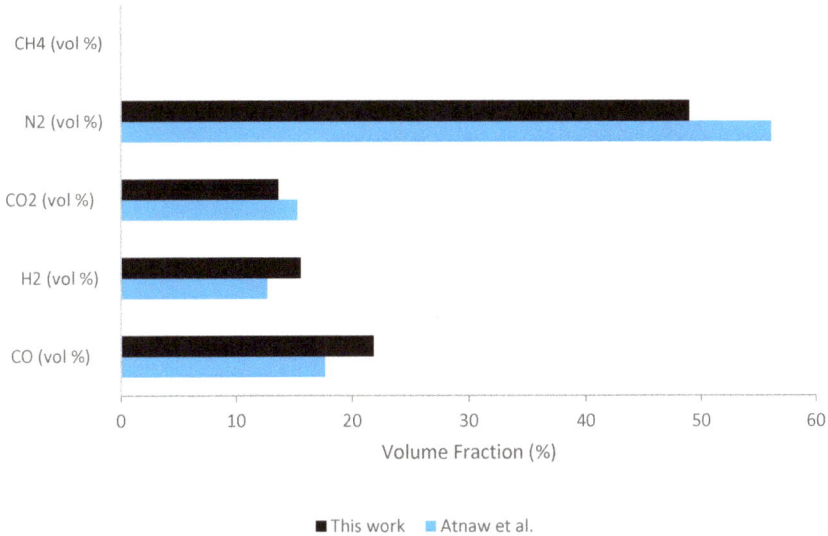

Figure 4. Syngas component yields predicted by this work's model (black line) versus the experimental work of Atnaw et al. (blue line).

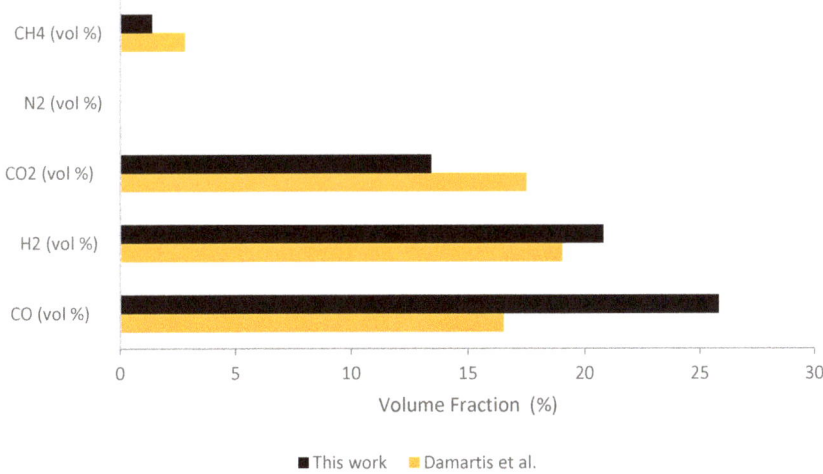

Figure 5. Syngas component yields predicted by this work's model (black line) versus the experimental work of Damartzis et al. (yellow line).

The accuracy of the simulation predictions was quantified using the root mean square deviation. According to this method, Root Sum Square (RSS) quantity is computed using the formula:

$$\text{RSS} = \sum ((y_{i,l} - y_{i,p})/y_{i,l})^2, \tag{5}$$

where $y_{i,l}$ is the literature value of the volume fraction of a specific syngas component and $y_{i,p}$ the predicted value of the proposed Aspen Plus model at the same conditions. Then, using the total number of data (N), the mean root sum square quantity is calculated via the equation:

$$MRSS = RSS/N. \tag{6}$$

The mean error is defined by the equation:

$$\text{Mean Error} = (MRSS)^{1/2}. \tag{7}$$

The mean errors between the proposed Aspen Plus simulation layout and the literature results are shown in Table 15. A maximum error up to 30–35% was calculated for the CO as well as for the H_2 yield. For the other syngas components, errors were in the range of 10–30%. Such relatively high errors are anticipated when equilibrium models are implemented [50,51]. Overall, it is shown that the developed model can reliably predict the lower heating value of the produced syngas and the cold gas efficiency of the process which would subsequently be used as input for the LCA approach.

Table 15. Mean error of simulation results, when compared to the experimental work of Atnaw et al. and Damartzis et al., computed by Equation (7).

Species	Atnaw et al. (2018) (Experimental) [53]	Damartzis et al. (2012) (Numerical) [17]
CO	30	35
H_2	29	29
CO_2	21	15
N_2	9	N/A
CH_4	N/A	31

3.1.2. Impact of Gasification Temperature and ER to the Cold Gas Efficiency

After the validation of the Aspen Plus model, a sensitivity analysis was carried out to investigate the effects of gasification temperature and equivalence ratio to the cold gas efficiency of the system. It involved the prediction of syngas component yields, the calculation of the lower heating value of the produced syngas and the estimation of the process cold gas efficiency for all examined feedstocks. Indicative results for the gasification of corn stover are presented in Figures 6 and 7. Similar trends were observed for all other feedstocks. It should be noted that gasification temperature corresponds to the reduction zone temperature, since the composition and the lower heating value of the produced syngas are mainly affected by reactions occurring at this stage of the gasification process [60].

In Figure 6, it is shown that the gasification temperature has a major impact on the predicted cold gas efficiency under a specified equivalence ratio. Specifically, as temperature rises, the cold gas efficiency sharply increases (up to temperatures of approximately 750–800 °C) and then levels out at a maximum value, which differs according to the constant equivalence ratio (for example 68% in the case of ER = 0.2). This behavior can be explained by the nature of the reduction zone reactions. In particular, the equilibrium of the endothermic reactions R-6, R-7 and R-9 and the exothermic R-5 and R-8 moves towards the production of CO and H_2. As a result, the produced syngas has a higher volume fraction of CO and H_2 and consequently a higher LHV (due to Equation (3)) and a higher yield of combustibles. Thus, the gasification efficiency increases.

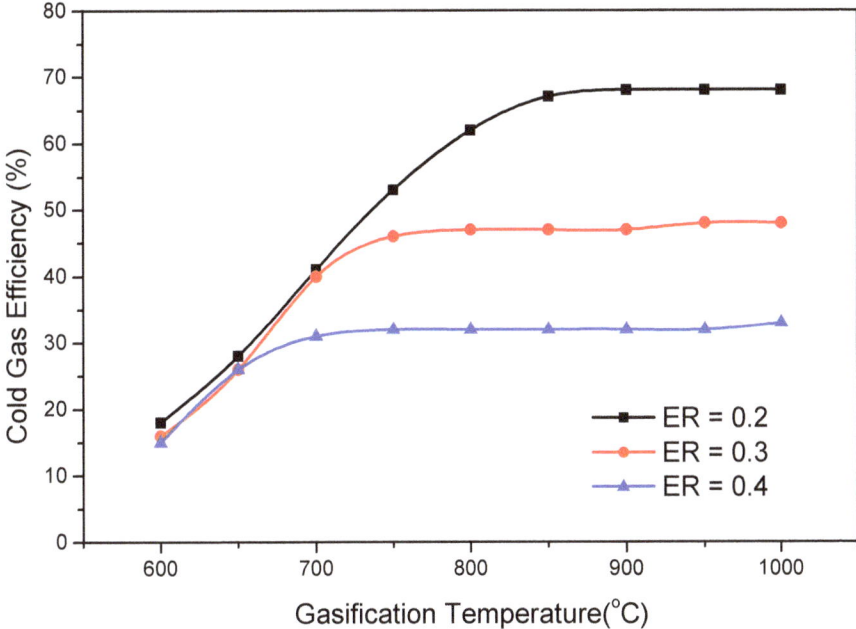

Figure 6. Effect of gasification temperature on the cold gas efficiency for the gasification of corn stover under the specified equivalence ratio (black line: ER = 0.2, red line: ER = 0.3, blue line: ER = 0.4).

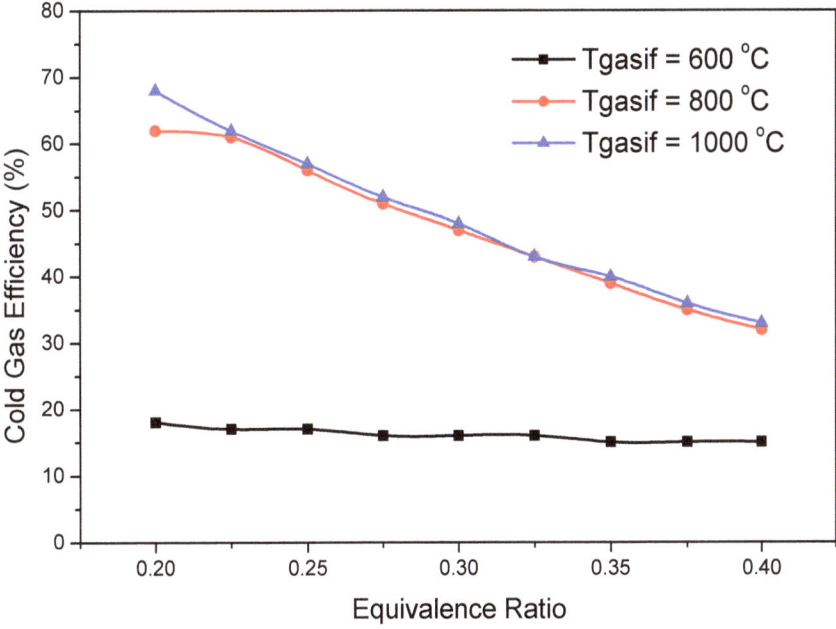

Figure 7. Effect of equivalence ratio to the cold gas efficiency for the gasification of corn stover under the specified gasification temperature values (black line: T_{gasif} = 600 °C, red line: T_{gasif} = 800 °C, blue line: T_{gasif} = 1000 °C).

Figure 7 presents the effect of the equivalence ratio on the cold gas efficiency of the gasification of corn stover at a specific gasification temperature. As an overall trend, the cold gas efficiency decreases as equivalence ratio shifts towards higher values. This can be associated with the promotion of complete oxidation conditions, and thus the decrease of the CO and H_2 yield in the produced syngas, as well as the increase of N_2 yield. As a consequence, the lower heating value and the volume flow of syngas decreases, lowering the gasification cold gas efficiency. Furthermore, for a gasification temperature of 600 °C, the cold gas efficiency is nearly constant at a minimum value of 16%, because, due to the low system temperature, the equilibria of reactions R5–R9 result to low and nearly constant CO and H_2 values. On the contrary, in case of a gasification temperature of 800 °C and 1000 °C, the respective equilibrium is significantly shifted towards high CO and H_2 values. It should be noted that the cold gas efficiency has not been calculated for equivalence ratios lower than 0.2, since a further reduction of the system's efficiency would be expected due to the promotion of complete pyrolysis conditions, which lower the overall heat transfer rate and thus decrease the volatile compounds that take part into gasification reactions [61,62].

In addition, according to Figure 6, the maximization of the cold gas efficiency of the gasification process occurs for an equivalence ratio value of 0.2. As a result, ER = 0.2 is considered the optimal gasification equivalence ratio for all feedstocks involved in this study. Figure 8 describes the effect of gasification temperature on the cold gas efficiency for the five feedstocks examined in this study, at ER = 0.2. As it is observed, efficiencies approximating or marginally surpassing 70% were predicted for the five feedstocks. Differences between predicted efficiencies of individual feedstocks are associated with the specific characteristics of their proximate and ultimate analyses.

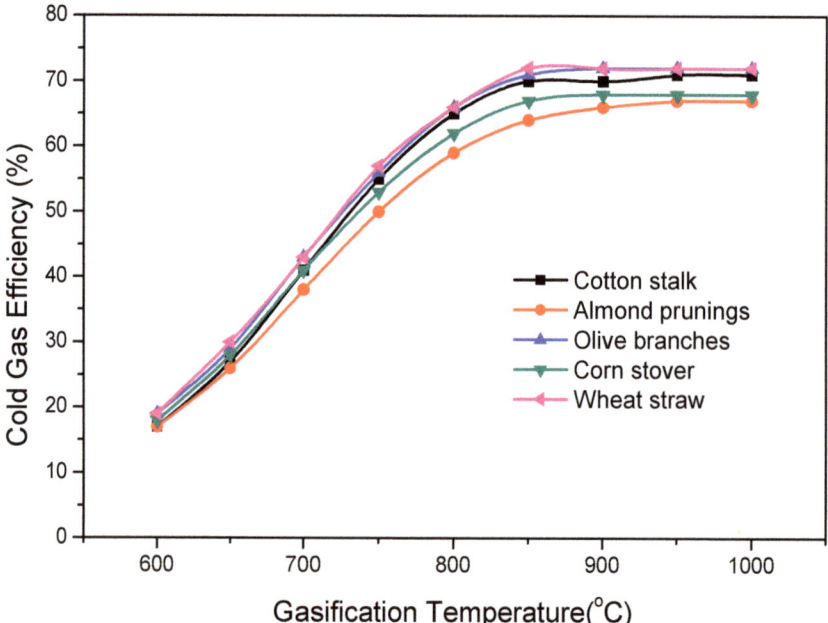

Figure 8. Effect of gasification temperature on the cold gas efficiency of the gasification process for every feedstock examined for ER = 0.2 (black line: cotton stalk, red line: almond prunings, blue line: olive branches, green line: corn stover, purple line: wheat straw).

The optimal operating points (i.e., those in which the cold gas efficiency is maximized) of the proposed gasification layout for each feedstock involved in this study are given in Table 16.

The maximum cold gas efficiency for each feedstock was used as input for the LCA study, describing the optimal conditions, in which the gasifier provides the required thermal power for the needs of the proposed 1 MW$_{el}$ and 2.25 MW$_{th}$ CHP plant.

Table 16. Optimal operating points of the proposed gasification layout for each feedstock examined.

Feedstock	Gasification Temperature (°C)	Equivalence Ratio	Syngas LHV (MJ/m^3)	CGE (%)
Cotton stalk	850	0.2	7.38	70
Corn stover	850	0.2	7.33	67
Olive branches	850	0.2	7.58	71
Almond prunings	900	0.2	7.53	66
Wheat straw	850	0.2	7.44	72

The calculated optimal cold gas efficiencies do not significantly deviate from the respective values reported for the Güssing CHP plant (ranging between 60 and 70%, according to [37]. Given that the Güssing power plant utilizes a steam blown fluidized bed gasifier instead of the fixed bed gasifier considered in this work, this relative agreement of cold gas efficiency values enabled us to adopt the Güssing CHP plant data as input in the LCA calculations without anticipating notable discrepancies.

3.1.3. Impact of Initial Biomass Moisture Content on Cold Gas Efficiency

The effect of the initial moisture content of each feedstock on the cold gas efficiency was investigated using the developed Aspen Plus model and was used as input for the LCA study. Simulations were performed at an equivalence ratio of 0.2 and for raw biomass moisture content ranging from 0 to 30%. The drying process was assumed to be complete, so the final biomass moisture content was 0%. Figure 9 shows the relationship between cold gas efficiency and initial moisture content for each feedstock examined in this study. As initial moisture content rises, more heat is required for the drying process, which lowers the gasification temperature and affects the reduction zone reactions, resulting in lower CO and H$_2$ yields. Consequently, the syngas LHV decreases and the cold gas efficiency shows a relative decrease of approximately 35% for all feedstocks examined.

A crucial factor for the LCA study was the investigation of whether the gasification process was energetically self-sufficient. As described before, the drying process consumes a lot of thermal energy, which not only lowers the gasification efficiency but may require extra heat by, e.g., fossil fuel combustion. It should be made clear whether additional heat is required or not, in order to adapt the LCA model. Towards clarifying this issue, it was initially assumed that biomass drying is done by a heat exchanger which used the rejected heat from the syngas cooler. As stated by Rentizelas et al. (2008), corn stover has the maximum initial moisture content, which is 50%. Using the Aspen Plus simulation modules, the heat required for biomass drying and the heat released from syngas cooling were calculated per kg of biomass feed for an initial biomass moisture content of up to 50%. The results of this investigation are presented in Figure 10. It is clear not only that the heat derived from syngas cooling is sufficient for biomass drying, even in the unfavorable scenario examined, but also that extra heat can be used for other purposes, for example in the boiler which produces thermal power in the CHP plant.

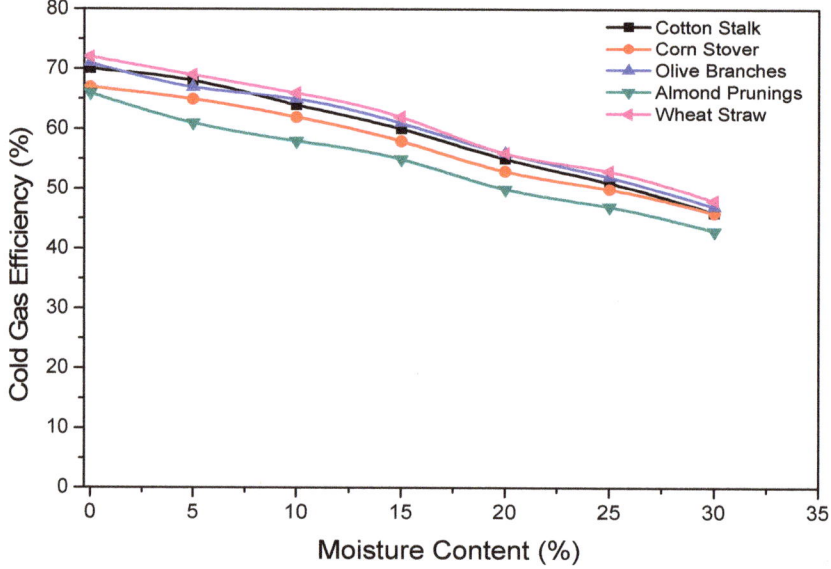

Figure 9. Effect of Moisture Content of the biomass types examined on the Cold Gas Efficiency of the gasification process (black line: cotton stalk, red line: almond prunings, blue line: olive branches, green line: corn stover, purple line: wheat straw).

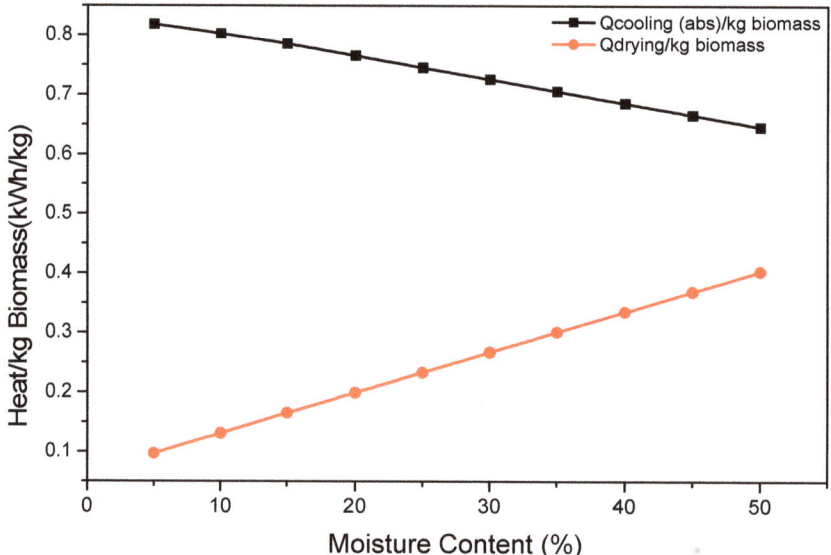

Figure 10. Heat required for biomass drying and heat rejected by the syngas cooling per kg of biomass feed in the case of corn stover.

3.2. Life Cycle Assessment Results

3.2.1. CHP Plant Environmental Hotspots and Comparison with the Greek Natural Gas Supply Chain

One of the scopes of this work is to highlight the operational parameters of the CHP biomass gasification plant which has the major share in the examined impact categories. In this analysis, only feedstocks that are transported to the plant were examined, since transportation enlarges the environmental burdens of the total process. Moreover, by assuming the same transport distance for all feedstocks involved in this study, the environmental hotspot results were expected to follow the same trend for all biomass types. So, results for only one random feedstock (almond prunings) are presented in this section. Figures 11 and 12 show the impact assessment results for the system environmental hotspot as well as the comparison of the syngas and natural gas supply chains. It should be noted that both impact categories were assessed per MJ of gas energy because syngas and natural gas lower heating values are significantly different.

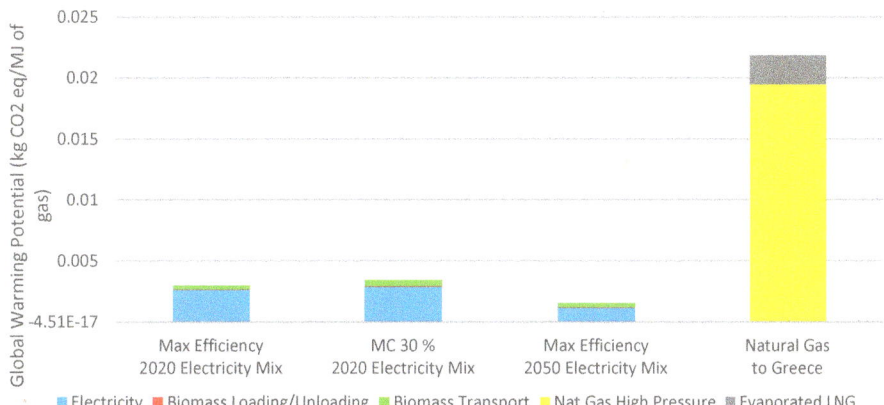

Figure 11. CHP plant environmental hotspot analysis and comparison of syngas and natural gas supply chain regarding the Global Warming Potential impact category (units: kgCO$_2$ eq/MJ of gas).

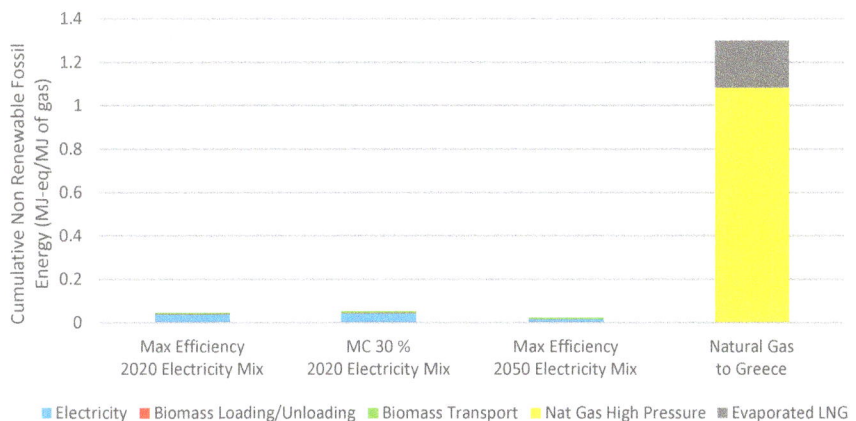

Figure 12. CHP plant environmental hotspot analysis and comparison of syngas and natural gas supply chain regarding the Cumulative Energy Demand of Non-Renewable Fossil Energy impact category (units: MJ of fossil energy/MJ of gas).

In terms of the Global Warming Potential and Cumulative Energy Demand of Non-Renewable Fossil Energy impact categories, the parasitic load of the plant was determined to be the system environmental hotspot, contributing to approximately 90% of the total environmental burdens of the plant in all moisture content and electricity mixture scenarios examined. This is explained by the fact that the CHP plant uses electricity directly from the grid. More specifically, in the current Greek electricity mixture as well as the projection for 2050 under current policies, energy production from fossil fuels is included, which enlarges the environmental burdens of the plant. However, the total environmental burdens decrease about 50% when 2050 electricity mixture is supplied, due to the higher share of renewables in the mix.

Furthermore, the comparison of the aforementioned impact categories for the production of 1 MJ of energy via syngas and natural gas burn quantifies the environmental benefits of the use of syngas instead of natural gas. In detail, syngas exploitation, instead of natural gas consumption, contributed to a large reduction in both impact categories assessed, even if the plant operated at adversary conditions (30% initial biomass moisture content). GHG emissions and fossil fuel use become even lower when the 2050 current policy electricity mixture is concerned. These results are explained by the fact that the Greek natural gas supply is responsible for considerable emissions during extraction and pipeline transportation, which enlarge the corresponding environmental burdens.

3.2.2. CHP Plant Comparison with Conventional Reference Cases

In order to highlight the environmental benefits of the CHP biomass gasification plant simulated in this work, its environmental footprint should be compared with conventional energy production alternatives. The analysis involved the plant operation at optimal (max efficiency) and most unfavorable (30% initial feedstock moisture content) condition, as well as operation under the 2020 and the 2050 electricity generation mixtures. Figures 13 and 14 summarize the Life Cycle Assessment results for the production of 1 kWh of electricity by the CHP biomass gasification plant and the reference cases examined. It should be noted that the negative columns represent the co-generated heat of the CHP plants, which, based on the ISO 14040 standard, was included in the study as an avoided product (i.e., the corresponding operation of a typical industrial gas boiler is avoided). Plant emissions and fossil energy demand were calculated as the sum of the positive and the negative columns and are given upon the bars. Negative sum values of the indicators considered in this study mean that the operation of the plant leads to GHG mitigation and fossil energy savings.

The LCA results presented in Figures 13 and 14 show that the operation of the CHP biomass gasification plant, under all conditions examined, leads to GHG mitigation (approximately 0.6 kg CO_2eq per kWh_{el}) and non-renewable energy savings (approximately 10 MJ per kWh_{el}). This finding is justified by the assumption of a) assigning zero burden to the biomass growth stage (agricultural waste) and b) zero contribution to climate change from biogenic CO_2 emissions. A quite significant outcome of the LCA study was that the CHP biomass plant operation, under all operating conditions, can lead to CO_2 mitigation and fossil energy savings which are nearly equal to the emissions and the fossil fuel use for production of the same amount of electricity from the 2020 Greek energy scheme. Furthermore, the considered power plant can be environmentally beneficial, even when compared to the kWh generated by the envisaged 2050 Greek electricity mixture, which includes a larger share of renewables. Finally, the biomass gasification plant had clearly less impact on climate change than the natural gas internal combustion engine on CHP mode, due to the emissions and the energy use associated with natural gas supply and use.

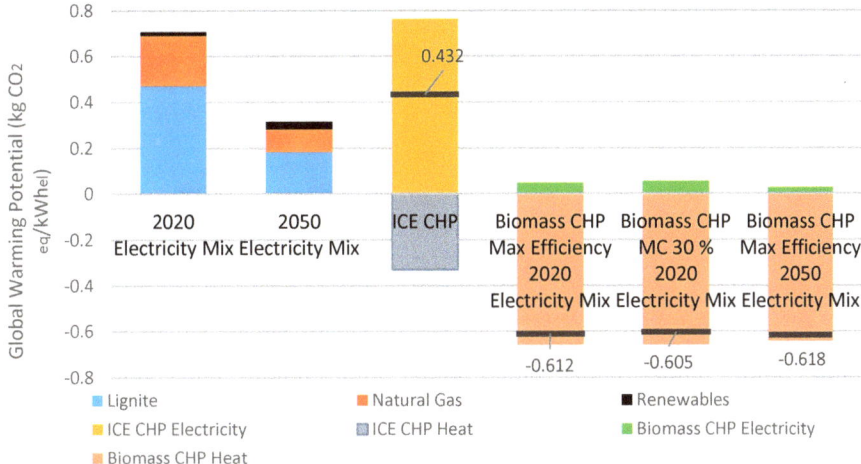

Figure 13. Global Warming Potential impact category results for the simulated CHP biomass gasification plant at different operating conditions, and the conventional energy production alternatives (electricity from the 2020 and 2050 grid, natural gas internal combustion engine on CHP mode). Units: kg CO_2eq/ kWh$_{el}$.

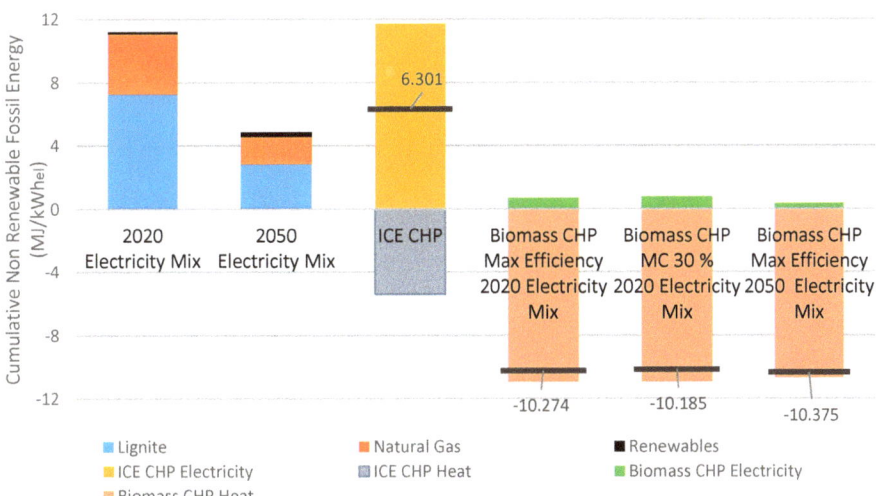

Figure 14. Cumulative demand of non-renewable fossil energy impact category results for the simulated CHP biomass gasification plant at different operating conditions, and the conventional energy production alternatives (electricity from the 2020 and 2050 grid, natural gas internal combustion engine on CHP mode). Units: MJ/ kWh$_{el}$.

The results presented in Figures 13 and 14 show a vast environmental advantage of the kWh generated from syngas, but there are two critical parameters whose influence must be evaluated: (a) the percentage of biomass CHP heat utilization and (b) the annual variation of biomass availability. Regarding the first parameter, the results calculated so far assume that all the co-generated heat will be used (replacing the heat from fossil fuel combustion), but this assumption can be considered as over-optimistic. Therefore, a "zero-credit from CHP heat utilization" case was be examined (0% heat use), in order to facilitate a realistic situation where a partial utilization takes place. Feedstock availability can fluctuate considerably, according to climatic or market influences. The variable biomass input has a straightforward effect on the annual working hours of the CHP biomass gasification plant. Using the feedstock availability figures, which were presented on Table 4, multi and single feedstock operations were < considered. In the multi feedstock operation, biomass quantities were sufficient for full year operation (8760 h) and 8.76 GWh$_{el}$ were annually produced. In the single feedstock operation, the power plant was able to use only in-house cotton stalk as fuel. Thus, the annual working hours were drastically reduced to 2535, which corresponded to 2.535 GWh of electricity per year.

The results showing the influence of the aforementioned parameters are presented in Figures 15 and 16, where the CHP biomass gasification plant is compared to conventional energy production alternatives in terms of GHG mitigation and fossil energy savings per year. The annual emissions and energy use of operation of the CHP biomass gasification plant are compared with those from the production of electricity from the 2020 and the 2050 Greek electricity mixture, as well as from the natural gas internal combustion engine at CHP mode.

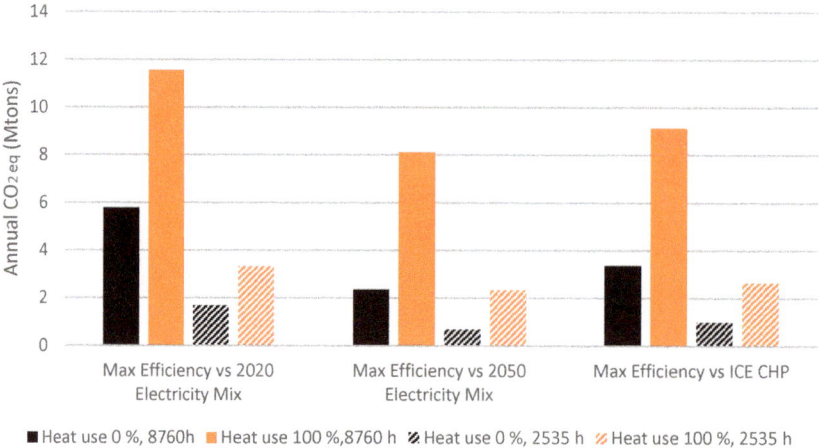

Figure 15. Annual GHG mitigation from the CHP biomass gasification plant considered in this study when compared with electricity production from the Greek 2020 and 2050 mix and with a natural gas internal combustion engine on CHP mode (Units: Annual Mtons of CO_2eq).

On the bright side, all parametric cases resulted in a better biomass CHP performance, both in terms of emissions and non-renewable energy demand. However, the advantage of biomass CHP was drastically reduced. If the negative effect of both parameters is considered, the annual CO_2 mitigation (Figure 15) and non-renewable energy savings (Figure 16) were reduced by a factor of 6 to 9, depending on the comparison. The biomass CHP advantages were reduced by a factor of 2 to 4 if only the zero heat utilization credit case is calculated. The corresponding reducing factor of low biomass availability lay between 3 and 4. Therefore, maximizing both the CHP heat utilization and the plant annual operation should be targeted.

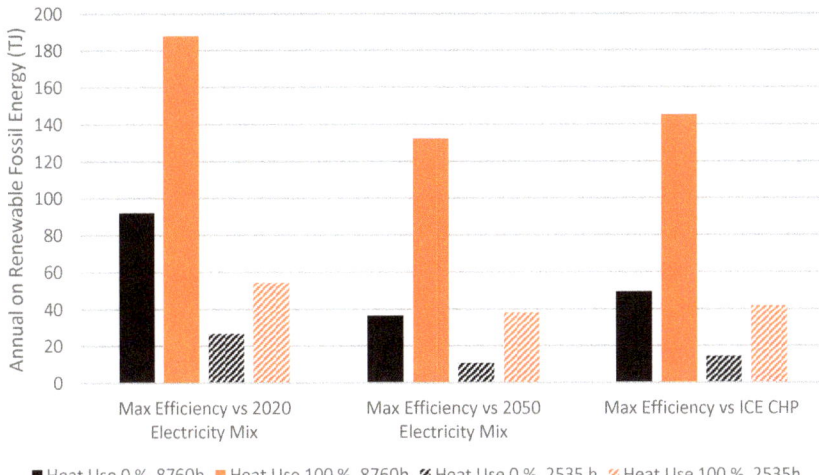

Figure 16. Annual fossil energy savings from the CHP biomass gasification plant considered in this study when compared with electricity production from the Greek 2020 and 2050 mix and with a natural gas internal combustion engine on CHP mode (Units: TJ of fossil energy).

4. Conclusions

This study aims at assessing the energetic and environmental performance of a prospective cogeneration biomass gasification plant situated in Thessaly, Greece, via a combined process simulation and the Life Cycle Assessment method. Initially, the basic operational parameters of the prospective 1 MW_{el} CHP biomass gasification plant were obtained from the literature. The most common agricultural residues in Thessaly, Greece, were identified, and the contribution of each biomass type to the total annual feedstock demand was determined.

The developed equilibrium process model quantified the effect of gasification temperature, equivalence ratio and raw biomass moisture content on the gasification of the examined feedstock types. The modeling approach was validated by comparing the predicted syngas LHV and syngas species yields against measured values from the literature, with maximum deviations in the predicted LHV in the range of 10–15%. Simulations of the biomass gasification process revealed a maximum gasification efficiency of approximately 70% for all examined feedstock types at ER = 0.2, while lower efficiency values were observed when the raw biomass moisture content increased.

After upscaling the gasification model to a 1 MW_{el} and 2.25 MW_{th} CHP plant, a "Cradle to Gate" Life Cycle Assessment was conducted and examined the Global Warming Potential and the Cumulative Demand of Non-Renewable Fossil Energy of the prospective plant. Provided that zero burden is to be assigned to the biomass growth stage (being agricultural waste) and that zero contribution is to be considered for climate change from biogenic CO_2 emissions, results identify the plant electricity consumption as the main plant environmental hotspot. The results suggest that plant operation in all examined conditions leads to GHG mitigation and non-renewable energy savings of approximately 0.6 kg CO_2eq/kWh_{el} and 10 MJ/kWh_{el}, respectively. Nevertheless, the advantage of biomass CHP is considerably affected by the negative effects of the percentage of biomass CHP heat utilization and the annual variation of biomass availability. Within a context of zero CHP heat utilization and minimum feedstock availability, the annual CO_2 mitigation and non-renewable energy savings are reduced by a factor of 6 to 9, depending on the comparison.

Author Contributions: Conceptualization, D.K., D.G. and M.F.; methodology, I.V., D.K and D.G.; software, I.V.; validation, D.K. and D.G.; formal analysis, I.V.; investigation, all authors; resources, M.F.; writing—original draft preparation, I.V.; writing—review and editing, D.K., D.G. and M.F.; visualization, I.V.; supervision, M.F.; project administration, D.K., D.G. and M.F.; funding acquisition, M.F. All authors have read and agreed to the published version of the manuscript.

Funding: This research has been co-financed by the European Union and Greek national funds through the Operational Program Competitiveness, Entrepreneurship and Innovation, under the call RESEARCH–CREATE–INNOVATE (project code: T1EDK-05563, project title: Three stage gasifier for power production from biomass).

Conflicts of Interest: The authors declare no conflict of interest.

References

1. European Union. Directive (EU) 2018/2001 of the European Parliament and of the Council on the promotion of the use of energy from renewable sources. *Off. J. Eur. Union* **2018**, *2018*, 1–128. Available online: https://eur-lex.europa.eu/eli/dir/2018/2001/oj (accessed on 31 July 2020).
2. European Commission. A Clean Planet for all—A European Strategic Long-Term Vision for a Prosperous Modern, Competitive and Climate Neutral Economy COM/2018/773 Final. Available online: https://eur-lex.europa.eu/legal-content/EN/TXT/?uri=CELEX:52018DC0773 (accessed on 18 September 2020).
3. Fuel Cells and Hydrogen Joint Undertaking (FCH). *Hydrogen Roadmap Europe: A Sustainable Pathway for the European Energy Transition*; FCH: Brussels, Belgium, 2019.
4. PricewaterhouseCoopers et al. Sustainable and Optimal Use of Biomass for Energy in the EU beyond 2020. VITO, Utrecht University, TU Vienna, INFRO, Rütter Soceco & PwC. 2017. Available online: https://ec.europa.eu/energy/sites/ener/files/documents/biosustain_annexes_final.pdf%0Ahttps://ec.europa.eu/energy/en/studies/sustainable-and-optimal-use-biomass-energy-eu-beyond-2020 (accessed on 31 July 2020).
5. van Eijk, R.J. Options for Increased Utilization of Ash from Biomass Combustion and Co-Firing. IEA Bioenergy Task 32 (Biomass Combustion) Deliverable D4. 2012, pp. 1–40. Available online: http://task32.ieabioenergy.com/wp-content/uploads/2017/03/Ash_Utilization_KEMA.pdf (accessed on 31 July 2020).
6. Livingston, W.R. Biomass Ash Characteristics and Behaviour in Combustion, Gasification and Pyrolysis Systems. Doosan Babcock Energy. 2007, p. 69. Available online: https://antioligarch.files.wordpress.com/2014/12/biomass-fly-ash-characteristics-behaviour-in-combustion.pdf (accessed on 31 July 2020).
7. van Alkemade, M.M.C.; Loo, S.; Sulilatu, W.F. Exploratory investigation into the possibilities of processing ash produced in the combustion of reject wood. *Netherl. Organ. Appl. Sci. Res. Apeldoorn* **1999**, 1–45. Available online: http://task32.ieabioenergy.com/wp-content/uploads/2017/03/R99357.pdf (accessed on 31 July 2020).
8. Baruah, D.; Baruah, D.C. Modeling of biomass gasification: A review. *Renew. Sustain. Energy Rev.* **2014**, *39*, 806–815. [CrossRef]
9. Gambarotta, A.; Morini, M.; Zubani, A. A non-stoichiometric equilibrium model for the simulation of the biomass gasification process. *Appl. Energy* **2018**, *227*, 119–127. [CrossRef]
10. Basu, P. *Biomass Gasification and Pyrolysis Practical Design*; Academic Press: London, UK, 2010.
11. Breault, R.W. Gasification processes old and new: A basic review of the major technologies. *Energies* **2010**, *3*, 216–240. [CrossRef]
12. Kaupp, A.; Goss, J.R. *State of The Art Report for Small Scale (To 50 Kw) Gas Producer-Engine Systems*; PB-85-102002/XAD; Department of Agricultural Engineering, California University: Davis, CA, USA, 1981.
13. Martínez, L.V.; Rubiano, J.E.; Figueredo, M.; Gómez, M.F. Experimental study on the performance of gasification of corncobs in a downdraft fixed bed gasifier at various conditions. *Renew. Energy* **2020**, *148*, 1216–1226. [CrossRef]
14. Hai, I.U.; Sher, F.; Yaqoob, A.; Liu, H. Assessment of biomass energy potential for SRC willow woodchips in a pilot scale bubbling fluidized bed gasifier. *Fuel* **2019**, *258*, 116143. [CrossRef]
15. Yan, W.C.; Shen, Y.; You, S.; Sim, S.H.; Luo, Z.H.; Tong, Y.W.; Wang, C.-H. Model-based downdraft biomass gasifier operation and design for synthetic gas production. *J. Clean. Prod.* **2018**, *178*, 476–493. [CrossRef]

16. González, W.A.; Pérez, J.F.; Chapela, S.; Porteiro, J. Numerical analysis of wood biomass packing factor in a fixed-bed gasification process. *Renew. Energy* **2018**, *121*, 579–589. [CrossRef]
17. Damartzis, T.; Michailos, S.; Zabaniotou, A. Energetic assessment of a combined heat and power integrated biomass gasification-internal combustion engine system by using Aspen Plus®. *Fuel Process. Technol.* **2012**, *95*, 37–44. [CrossRef]
18. Lan, W.; Chen, G.; Zhu, X.; Wang, X.; Liu, C.; Xu, B. Biomass gasification-gas turbine combustion for power generation system model based on ASPEN PLUS. *Sci. Total Environ.* **2018**, *628–629*, 1278–1286. [CrossRef] [PubMed]
19. Han, J.; Liang, Y.; Hu, J.; Qin, L.; Street, J.; Lu, Y.; Yu, F. Modeling downdraft biomass gasification process by restricting chemical reaction equilibrium with Aspen Plus. *Energy Convers. Manag.* **2017**, *153*, 641–648. [CrossRef]
20. Wei, L.; Thomasson, J.A.; Bricka, R.M.; Sui, R.; Wooten, J.R.; Columbus, E.P. Syn-Gas Quality Evaluation for Biomass Gasification with a Downdraft Gasifier. *Trans. ASABE* **2009**, *52*, 21–37. [CrossRef]
21. Marcantonio, V.; Bocci, E.; Monarca, D. Development of a chemical quasi-equilibrium model of biomass waste gasification in a fluidized-bed reactor by using Aspen plus. *Energies* **2019**, *13*, 53. [CrossRef]
22. Navas-Anguita, Z.; Cruz, P.L.; Martín-Gamboa, M.; Iribarren, D.; Dufour, J. Simulation and life cycle assessment of synthetic fuels produced via biogas dry reforming and Fischer-Tropsch synthesis. *Fuel* **2019**, *235*, 1492–1500. [CrossRef]
23. The International Standards Organisation (ISO). *ISO 14040:2006 Environmental Management—Life Cycle Assessment—Principles and Framework*, 2nd ed.; ISO: Geneva, Switzerland, 2006.
24. The International Standards Organisation (ISO). International Standard Assessment—Requirements and guilelines. *Int. J. Life Cycle Assess* **2006**, 652–668. [CrossRef]
25. Adams, P.W.R.; McManus, M.C. Small-scale biomass gasification CHP utilisation in industry: Energy and environmental evaluation. *Sustain. Energy Technol. Assess.* **2014**, *6*, 129–140. [CrossRef]
26. Kimminga, M.; Sundberga, C.; Nordberg, Å.; Baky, A.; Bernessona, S.; Norénb, O.; Hanssona, P.-A. Biomass from agriculture in small-scale combined heat and power plants—A comparative life cycle assessment. *Biomass Bioenergy* **2011**, *35*, 1572–1581. [CrossRef]
27. Yang, Q.; Zhou, H.; Zhang, X.; Nielsen, C.P.; Lia, J.; Lu, X.; Yanga, H.; Chen, H. Hybrid life-cycle assessment for energy consumption and greenhouse gas emissions of a typical biomass gasification power plant in China. *J. Clean. Prod.* **2018**, *205*, 661–671. [CrossRef]
28. Tagliaferri, C.; Evangelisti, S.; Clift, R.; Lettieri, P. Life cycle assessment of a biomass CHP plant in UK: The Heathrow energy centre case. *Chem. Eng. Res. Des.* **2018**, *133*, 210–221. [CrossRef]
29. Nguyen, T.L.T.; Hermansen, J.E. Life cycle environmental performance of miscanthus gasification versus other technologies for electricity production. *Sustain. Energy Technol. Assess.* **2015**, *9*, 81–94. [CrossRef]
30. Guerra, J.P.; Cardoso, F.H.; Nogueira, A.; Kulay, L. Thermodynamic and environmental analysis of scaling up cogeneration units driven by sugarcane biomass to enhance power exports. *Energies* **2018**, *11*, 73. [CrossRef]
31. García, C.A.; Morales, M.; Quintero, J.; Aroca, G.; Cardona, C.A. Environmental assessment of hydrogen production based on Pinus patula plantations in Colombia. *Energy* **2017**, *139*, 606–616. [CrossRef]
32. Suwatthikul, A.; Limprachaya, S.; Kittisupakorn, P.; Mujtaba, I.M. Simulation of steam gasification in a fluidized bed reactor with energy self-sufficient condition. *Energies* **2017**, *10*, 314. [CrossRef]
33. Hamedani, S.R.; Villarini, M.; Colantoni, A.; Moretti, M.; Bocci, E. Life cycle performance of hydrogen production via agro-industrial residue gasification-a small scale power plant study. *Energies* **2018**, *11*, 675. [CrossRef]
34. Hamedani, S.R.; del Zotto, L.; Bocci, E.; Colantoni, A.; Villarini, M. Eco-efficiency assessment of bioelectricity production from Iranian vineyard biomass gasification. *Biomass Bioenergy* **2019**, *127*, 105271. [CrossRef]
35. Bisht, A.S.; Thakur, N.S. Small scale biomass gasification plants for electricity generation in India: Resources, installation, technical aspects, sustainability criteria & policy. *Renew. Energy Focus* **2019**, *28*, 112–126. [CrossRef]
36. Cambero, C.; Sowlati, T. Assessment and optimization of forest biomass supply chains from economic, social and environmental perspectives—A review of literature. *Renew. Sustain. Energy Rev.* **2014**, *36*, 62–73. [CrossRef]

37. Rauch, R.; Hofbauer, H.; Bosch, K.; Siefert, I.; Aichernig, C.; Tremmel, H.; Voigtlaender, K.; Koch, R.; Lehner, R. Steam Gasification of Biomass at CHP Plant Guessing–Status of the Demonstration Plant. In Proceedings of the 2nd World Conference on Biomass for Energy, Industry and Climate Protection, Rome, Italy, 10–14 May 2004; pp. 1687–1690.
38. Rentizelas, A.A.; Tatsiopoulos, I.P.; Tolis, A. An optimization model for multi-biomass tri-generation energy supply. *Biomass Bioenergy* **2009**, *33*, 223–233. [CrossRef]
39. Voivontas, D.; Assimacopoulos, D.; Koukios, E.G. Assessment of Biomass Potential for Power Production: A GIS Based Method. *Biomass and Bioenergy* **2001**, *20*, 101–112. [CrossRef]
40. Papadopoulos, D.P.; Katsigiannis, P.A. Biomass energy surveying and techno-economic assessment of suitable CHP system installations. *Biomass Bioenergy* **2002**, *22*, 105–124. [CrossRef]
41. Phyllis2—Database for the Physico-Chemical Composition of (Treated) Lignocellulosic Biomass, Micro- and Macroalgae, Various Feedstocks for Biogas Production and Biochar. Available online: https://phyllis.nl/ (accessed on 8 December 2019).
42. Pütün, A.E.; Özbay, N.; Önal, E.P.; Pütün, E. Fixed-bed pyrolysis of cotton stalk for liquid and solid products. *Fuel Process Technol.* **2005**, *86*, 1207–1219. [CrossRef]
43. Klimantos, P.; Koukouzas, N.; Katsiadakis, A.; Kakaras, E. Air-blown biomass gasification combined cycles (BGCC): System analysis and economic assessment. *Energy* **2009**, *34*, 708–714. [CrossRef]
44. Di Blasi, C.; Branca, C. Modeling a stratified downdraft wood gasifier with primary and secondary air entry. *Fuel* **2013**, *104*, 847–860. [CrossRef]
45. Bhattacharya, S.C.; Hla, S.S.; Pham, H. A study on a multi-stage hybrid gasifyer-engine system. *Biomass Bioenergy* **2001**, *21*, 445–460. [CrossRef]
46. Tavares, R.; Monteiro, E.; Tabet, F.; Rouboa, A. Numerical investigation of optimum operating conditions for syngas and hydrogen production from biomass gasification using Aspen Plus. *Renew. Energy* **2020**, *146*, 1309–1314. [CrossRef]
47. Olgun, H.; Ozdogan, S.; Yinesor, G. Results with a bench scale downdraft biomass gasifier for agricultural and forestry residues. *Biomass Bioenergy* **2011**, *35*, 572–580. [CrossRef]
48. Kakaras, E.; Karellas, S. Pollution Abatement Technology for Thermal Plants. Tsotras 2013. Available online: https://www.researchgate.net/publication/321914527_Pollution_Abatement_Technology_for_thermal_plants (accessed on 31 July 2020). (In Greek).
49. Lv, P.M.; Xiong, Z.H.; Chang, J.; Wu, C.Z.; Chen, Y.; Zhu, J.X. An experimental study on biomass air-steam gasification in a fluidized bed. *Bioresour. Technol.* **2004**, *95*, 95–101. [CrossRef]
50. Formica, M.; Frigo, S.; Gabbrielli, R. Development of a new steady state zero-dimensional simulation model for woody biomass gasification in a full scale plant. *Energy Convers. Manag.* **2016**, *120*, 358–369. [CrossRef]
51. Gagliano, A.; Nocera, F.; Bruno, M.; Cardillo, G. Development of an Equilibrium-based Model of Gasification of Biomass by Aspen Plus. *Energy Procedia* **2017**, *111*, 1010–1019. [CrossRef]
52. Fernandez-Lopez, M.; Pedroche, J.; Valverde, J.L.; Sanchez-Silva, L. Simulation of the gasification of animal wastes in a dual gasifier using Aspen Plus®. *Energy Convers. Manag.* **2017**, *140*, 211–217. [CrossRef]
53. Atnaw, S.M.; Sulaiman, S.A.; Singh, L.; Wahid, Z.A.; Yahya, C.K.M.F.B.C.K. Modeling and parametric study for maximizing heating value of gasification syngas. *BioResources* **2017**, *12*, 2548–2564. [CrossRef]
54. Asadullah, M. Barriers of commercial power generation using biomass gasification gas: A review. *Renew. Sustain. Energy Rev.* **2014**, *29*, 201–215. [CrossRef]
55. Baumann, H.; Tillman, A.-M. *The Hitch Hiker's Guide to LCA*; Studentlitteratur AB: Lund, Sweden, 2004.
56. Ecoinvent-Version-2 @ www.ecoinvent.org. Available online: https://www.ecoinvent.org/database/older-versions/ecoinvent-version-2/ecoinvent-version-2.html (accessed on 31 July 2020).
57. Lindzen, R.S. Climate change, the IPCC Scientific Assessment. Edited by J. T. Houghton, G. J. Jenkins and J. J. Ephraums. Cambridge University Press. Pp. 365 + pp. 34 summary. Hardback £40.00, paperback £15.00. *Q. J. R. Meteorol. Soc.* **1991**, *117*, 651–652. [CrossRef]
58. Althaus, H.; Bauer, C.; Doka, G.; Dones, R.; Frischknecht, R.; Hellweg, S.; Humbert, S.; Jungbluth, N.; Köllner, T.; Loerincik, Y.; et al. *Implementation of Life Cycle Impact Assessment Methods Data v2.2 (2010)*; Hischier, R., Weidema, B., Eds.; Swiss Centre for Life Cycle Inventories: Dübendorf, Switzerland, 2010; Ecoinvent Report No. 3; Available online: https://www.ecoinvent.org/files/201007_hischier_weidema_implementation_of_lcia_methods.pdf (accessed on 31 July 2020).

59. Villarini, M.; Marcantonio, V.; Colantoni, A.; Bocci, E. Sensitivity analysis of different parameters on the performance of a CHP internal combustion engine system fed by a biomass waste gasifier. *Energies* **2019**, *12*, 688. [CrossRef]
60. Susastriawan, A.A.P.; Saptoadi, H. Purnomo Small-scale downdraft gasifiers for biomass gasification: A review. *Renew. Sustain. Energy Rev.* **2017**, *76*, 989–1003. [CrossRef]
61. Li, X.T.; Grace, J.R.; Lim, C.J.; Watkinson, A.P.; Chen, H.P.; Kim, J.R. Biomass gasification in a circulating fluidized bed. *Biomass Bioenergy* **2004**, *26*, 171–193. [CrossRef]
62. Reed, T.B.; Walt, R.; Ellis, S.; Das, A.; Deutch, S. Superficial Velocity—The Key to Downdraft Gasification. In Proceedings of the 4th Biomass Conference of the Americas, Oakland, CA, USA, 29 August–2 September 1999.

© 2020 by the authors. Licensee MDPI, Basel, Switzerland. This article is an open access article distributed under the terms and conditions of the Creative Commons Attribution (CC BY) license (http://creativecommons.org/licenses/by/4.0/).

Article

Kinetic Modelling of Biodegradability Data of Commercial Polymers Obtained under Aerobic Composting Conditions

Ilenia Rossetti [1,*], Francesco Conte [1] and Gianguido Ramis [2]

1. Chemical Plants and Industrial Chemistry Group, Dip. Chimica, Università degli Studi di Milano, CNR-SCITEC and INSTM Unit Milano-Università, Via C. Golgi 19, 20133 Milan, Italy; francesco.conte@unimi.it
2. Dip. Ing. Chimica, Civile ed Ambientale, Università degli Studi di Genova and INSTM Unit Genova, P.le Kennedy 1, 16129 Genoa, Italy; gianguidoramis@unige.it
* Correspondence: ilenia.rossetti@unimi.it

Abstract: Methods to treat kinetic data for the biodegradation of different plastic materials are comparatively discussed. Different samples of commercial formulates were tested for aerobic biodegradation in compost, following the standard ISO14855. Starting from the raw data, the conversion vs. time entries were elaborated using relatively simple kinetic models, such as integrated kinetic equations of zero, first and second order, through the Wilkinson model, or using a Michaelis Menten approach, which was previously reported in the literature. The results were validated against the experimental data and allowed for computation of the time for half degradation of the substrate and, by extrapolation, estimation of the final biodegradation time for all the materials tested. In particular, the Michaelis Menten approach fails in describing all the reported kinetics as well the zeroth- and second-order kinetics. The biodegradation pattern of one sample was described in detail through a simple first-order kinetics. By contrast, other substrates followed a more complex pathway, with rapid partial degradation, subsequently slowing. Therefore, a more conservative kinetic interpolation was needed. The different possible patterns are discussed, with a guide to the application of the most suitable kinetic model.

Keywords: kinetics; polyolefins; aerobic biodegradation; waste plastic materials; biodegradation of plastic; standard plastic testing

Citation: Rossetti, I.; Conte, F.; Ramis, G. Kinetic Modelling of Biodegradability Data of Commercial Polymers Obtained under Aerobic Composting Conditions. *Eng* 2021, 2, 54–68. https://doi.org/10.3390/eng2010005

Academic Editor: Antonio Gil Bravo

Received: 2 February 2021
Accepted: 18 February 2021
Published: 20 February 2021

Publisher's Note: MDPI stays neutral with regard to jurisdictional claims in published maps and institutional affiliations.

Copyright: © 2021 by the authors. Licensee MDPI, Basel, Switzerland. This article is an open access article distributed under the terms and conditions of the Creative Commons Attribution (CC BY) license (https://creativecommons.org/licenses/by/4.0/).

1. Introduction

Polymers are the most widely used materials in everyday life due to their wide availability, low cost, easy production and appreciable properties. In particular, they are widely used as packaging materials, especially polyolefins, because of their good mechanical properties, thermal stability, good barriers to carbon dioxide, oxygen and aromatic compounds [1]. The recycling cost of plastic packaging is often higher than that of producing virgin plastics, so several thousand tons of goods are dissipated into the environment every day, increasing the amount of waste products. This is a well-known problem, but public attention has grown due to accumulation in the water environment in the form of microplastics, interfering with biodiversity [2–4].

In recent years, several publications have appeared on the biodegradation process of polyolefins and there are many publications addressing the mechanism of their degradation, which is a strongly controversial issue. On the global scenario, the environmental impact of plastic residua is huge, with polyethylene (PE) and polypropylene (PP) predominating in waste disposal, since they are vastly used in food packaging. To a lower extent, polystyrene (PS) is less used but similarly recalcitrant to degradation compared to the other mentioned polyolefins. Overall, these materials are typically considered non-biodegradable, and various strategies are in place to improve their degradability [5].

Various biotic and abiotic routes were considered important, and investigations considered pure microorganisms and complex communities (marine environment, compost, etc.) [5,6]. Multi-microorganisms-based degradation, waxworms and fungi in general [7] were demonstrated to be more active in degradation than single bacterial colonies. The results were strongly affected by the size, shape, crystallinity (with amorphous polymeric isles being more degradable than highly crystalline ones), surface functionalisation, hydrophilicity, etc. Furthermore, pretreatment (e.g., UV) or additives may enhance the biodegradation rate, especially in cases of the addition of oxidizing catalysts (e.g., metal complexes) [8–10] and enzymes [5,11]. A very interesting summary of aerobic biodegradation tests on various polymeric substrates is reported in [12], evidencing very scattered results, likely due to their widely different formulations. Depending on the testing methods, some loss of material may also be due to erosion, but with the typically used standard procedures to evaluate biodegradability, the physical loss of the material is not admitted. More likely, the use of hybrid materials, such as degradable ones (starch) mixed with properly pretreated or promoted polyolefins, is an effective strategy to form (at least partially) a biodegradable material from what originally was a substantially non-biodegradable one. The need for a more standardized elaboration of the data derives from the very wide literature that addresses specific tests and approaches biodegradability over specific materials and looking to specific properties (tensile properties, surface functional groups, surface morphology, etc.).

Of all these works, summarized in very comprehensive reviews [5,9–15], only few propose testing according to standard conditions, and in each case the comparison between materials and conditions is hard, since absolute data are reported, e.g., y% degradation after x days. These are punctual data that help to semi-quantitatively assess the degradation behavior but do not suggest trends and do not allow any long-term prediction. As an example, a linear low-density polyethylene was formed as film with a high content of thermoplastic starch (TPS), leading to ca. 10 wt% loss after 5 months burial in soil [16]; the biodegradation of polystyrene in agricultural and desert soil was studied for a formulate with starch, possibly irradiated with gamma rays, evidencing maximum 10 wt% loss after 6 months burial [17]. This is important information for studying the behavior of the material in those conditions, but does not allow for understanding if a plateau is reached or if prolonging the burial will end in total biodegradation.

In summary, trends and mathematical predictive modelling are not provided. According to the latest literature data, a detailed kinetic study on the aerobic biodegradation of polymeric composites is not available and, in any case, kinetic data for commercial materials and suitable models to interpret them are substantially rare [18,19]. Hence, the scope of the work is to propose simple kinetic models for the interpretation of data collected through a standard biodegradation testing method. Such measures are forcedly limited in duration (a few weeks or months), but should be used to assess phenomena (final degradation) that are much broader in time. Thus, provided that data on the array of conversion vs. time data are available, we here compare different kinetic equations to interpret such data. It should be stated that the conclusion of this work is not to assess if one material is biodegradable or not, considering that the degradation of a formulate also strongly depends on the size of the substrate and possible unknown additives that were purposely added, but rather to understand the limits and features of the different kinetic models to approach the biodegradation phenomena. The availability of such information would allow for estimation of the time required for the final biodegradation (e.g., of 90%, 99% and 99.9%) of commercial polymers (or other materials) during aerobic degradation.

In order to cover this issue, experimental data for the aerobic biodegradation of three commercial materials were elaborated and collected according to a standard testing method. The materials were named A, B and C, without specific reference to the type of material. Indeed, a considerable debate is ongoing regarding the degradation behaviour of plastic materials, between those who support the possibility of degrading plastics, in particular, polyolefins, and skeptics. We decided not to focus on the nature of the tested materials in order to stress the primary meaning of this work, which is to comment on the features of very simple mathematical models that are useful to describe the degradation kinetics. We do not want to conclude or stress anything about the biodegradation potential of such materials, only to suggest the most correct methods to treat biodegradation data. Therefore, we have collected CO_2 evolution patterns vs. time for materials that showed different kinetics. This allowed for the proposal of a comparison between different models and to detail a very simple modelling procedure that, surprisingly, is rarely found in the considerable body of literature on this topic.

Different sets of data were compared and related to experiments of biodegradation according to a standard method ISO 14855 [11] on three commercial packaging formulates A, B and C, whose conversion vs. time pattern was markedly different. The results were also compared with cellulose as benchmark. In this work, we compared different kinetic models, which would allow for extrapolation of the ultimate biodegradation time. The approach consisted of first applying the simplest kinetic models (for example, integrated kinetic equations of order 0, 1, 2), checking the consistency of the previsions with the experimental data and then determining the estimated time of ultimate biodegradation. In some cases, it was necessary to apply more complex kinetic models, due to the inadequacy of the traditional ones.

2. Materials and Methods

The International Standard ISO 14855 is the method used to evaluate biodegradation, entitled "Determination of the final aerobic biodegradability of plastic materials under controlled condition of composting—Method by analysis of evolved carbon dioxide", and was designed for determining the biodegradability of plastic materials. This method is very accurate for determining the biodegradability with a small-scale vessel, involving evaluation of CO_2 trapped in absorption columns [20]. Different CO_2 testing methods have also been compared [15].

The method aims to simulate the typical composting conditions that occur during the disposal of urban solid wastes. A testing vessel is filled with a solid bed (compost, as in our case, or vermiculite), where the plastic material is dispersed, with initial mass m°, and added to an inoculum of stabilized mature compost deriving from the organic fraction of municipal wastes, which acts as a source of thermophilic microorganisms. Due to the difficult recovery of the residual solid, the method prescribes the determination of product conversion based on analysis of the CO_2 emitted from the cell. The reaction is monitored at given time intervals for no more than 6 months.

It should be underlined that the ISO 14855 standard protocol does not suggest the use or selection of specific microorganisms. Moreover, it does not prescribe any characterization of the compost. An internal reference is proposed, by comparing the degradation behavior of the plastic with a biodegradable reference. This notwithstanding, if this protocol is used to draw conclusions on the degradation mechanism or to assess the biodegradation of new materials, characterization of the compost and of the materials should be carried out, as reported, e.g., in [11].

The tests were carried out at 58 °C in a dark vessel (ca. 5 L). Aerobic conditions are ensured by CO_2-free air supply throughout the test (65% relative humidity, flowrate 100 mL/min). Outflowing CO_2 is sampled and determined by gas chromatographic analysis (Agilent, mod. 7890). The GC is equipped with first a HP Plot Q capillary column, followed by a molecular sieve (MS) one. Both columns were 0.53 mm, 30 m. The setting of a valve allows for exclusion of the MS. The assembly allows the quantification of CO_2, H_2O, O_2, N_2 and CO.

The studied materials are different commercial plastic formulates, named A, B and C, from food packaging. In every case, they were used in granular form, obtained by grinding and sieving with granules <1 mm in size, whose carbon content (C_{tot}) was determined by combustion and IR analysis of the CO_2 produced. Cellulose (thin-layer chromatography grade, <20 μm) was used both as a benchmark for a biodegradable material and to check the activity of the compost, as described in the standard procedure. No specific characterization of the compost and of the plastic material was carried out, since the aim of the work is to collect a CO_2 evolution vs. time dataset, rather than discuss the biodegradation behavior of the tested materials.

Three replicates were collected simultaneously for each sample under the same conditions. Testing was suspended after 180 days according to the norm prescriptions, even if incomplete conversion was achieved. The material was inspected after testing. Apparently, there were no significant residuals in the case of materials A, B and cellulose. Some residual grains were detected for material C, with surface deterioration, opacisation and striction.

The percentage of biodegradation of the material is calculated as the ratio between the CO_2 produced from the material at a given time, after subtraction of a blank sample (i.e., compost without plastic material) and the theoretical CO_2 amount (Th_{CO2}), corresponding to the mineralization of the whole sample [21]. This was calculated from Equation (1). An intrinsic unbalance is constituted by the use of some carbon to form the reacting biomass itself, which is not converted to CO_2 and determines the impossibility of reaching 100% conversion. Typically, this error is lower than 5–10%, but it is higher in some cases, and can be accounted for by comparing the data with a reference degradable material (here, cellulose).

The following calculations were done, starting from the raw data

$$Th_{CO2}\ (mg) = m°_{sample} \cdot C_{tot} \cdot \frac{44}{12} \quad (1)$$

$$\text{Conversion}\ (X) = \frac{m_{CO2}\ (\text{sample, t}) - m_{CO2}\ (\text{blank, t})}{Th_{CO2}} \quad (2)$$

where $m°_{sample}$ is the initial mass of the sample in mg, C_{tot} is the mass fraction of C in the sample and m_{CO2} represents the mass of CO_2 released at a given time from the sample vessel, to be compared with the one from a blank test (without sample). The data reported here represent the average between the three tests performed.

3. Results and Discussion

3.1. Material A

From the ratio between the amount of CO_2 produced at time t (raw data) and the theoretical one (corresponding to full biodegradation of the dry material mass loaded initially, m°, based on its carbon content, as Equation (1)), we calculated the fraction of CO_2 produced, which represents the degree of advancement (X) of the biodegradation reaction. The amount of residual polymer m(t) at time t was calculated from the formula

$$m(t) = m° \cdot (1 - X) \quad (3)$$

Figure 1 reports the conversion, as a percentage, as a function of time for the three tested samples of different materials and the cellulose used as reference.

The cellulose biodegradation data revealed very fast degradation in few days in the compost material used, assessing the validity of the test. It should be remembered that this test typically does not reach 100% conversion, since some carbon remains in the soil as part of the microorganism's metabolism. The last linear part of the curve of the reference cellulose is considered as full degradation being accomplished.

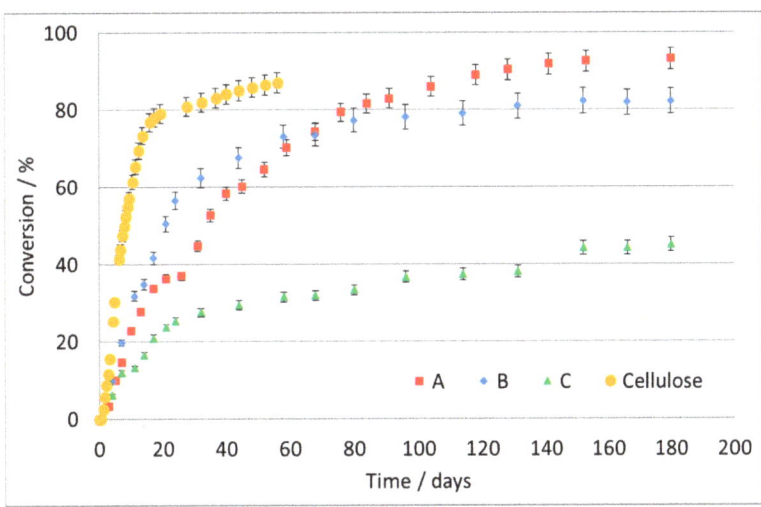

Figure 1. Conversion (%) of the three commercial samples and reference cellulose as a function of time.

These data were then processed by applying an integrated kinetic equation of the first order

$$\ln(m(t)/m°) = kt \quad (4)$$

where the time was expressed in days and k, the kinetic constant of the first order, expressed in days^{-1}. Data obtained using reaction orders of 0 or 2 with respect to the substrate are reported in the Supporting Information file.

The maximum conversion achieved for the A formulate was ca. 93% after 180 days. The degradation rate strongly depends on the material itself and on particle size: here, the particles were very small, with a high surface area, which facilitates the attack and colonization by bacteria, and is not directly comparable with plastic foils, films, etc. Furthermore, formulates were used where the presence of pro-oxidants cannot be excluded to improve the degradability. Besides dependence on the material, it is important to note the different testing methods, burial materials and temperatures in the different data reported in the literature.

It should be noted that is not possible to directly estimate the conversion time of the polymer at 100% because the CO_2 produced may not reach 100% compared with the theoretical, as stated above. In addition, it is noticed that a kinetic model of the first order does not admit the time estimate required for the conversion of 100% of the substrate, which would correspond to the computation of ln(0). We can then estimate the time required to achieve a conversion as close as desired to 100% (in our process: 90, 99, 99.9%).

Representing the results according to Equation (4), a substantially linear trend was observed, confirming the validity of the assumption of a first-order kinetics, except for the last points of the test, in which the reaction was substantially complete. Figure 2 shows the results obtained from the linear regression of the curve according to the first-order model for the A formulate. The last three points, marked in blue, were considered outliers according to a statistical analysis of the deviance of their error (>300% in absolute value), from the mean value of the difference between the value calculated by the regression and the experimental one.

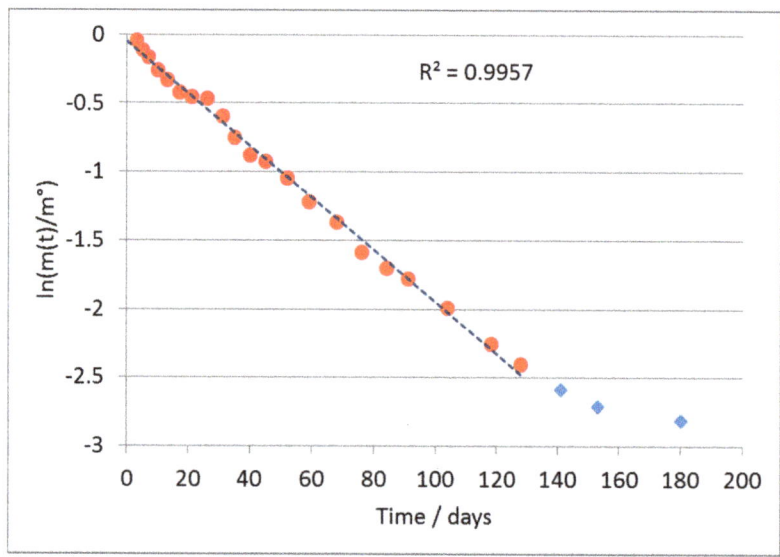

Figure 2. Linear regression of the conversion data for the A formulate according to the first-order model (Equation (4)). Blue diamonds were not included in the regression pertaining to the plateau region.

The absolute value of the slope of the line directly provides the value of the kinetic constant k, according to Equation (4). It was calculated as a value of $k = 0.0204 \pm 0.0010$ days^{-1}.

As verification, we calculated the half-life of the polymer, $t_{1/2}$. For kinetics of first order, the calculation of the half-life is carried out with the following equation

$$t_{1/2} = \ln(2)/k = 34 \text{ days} \tag{5}$$

This result is perfectly in line with the experimental data, thus confirming the correct attribution of the reaction order. Then, the time required to biodegrade the polymer to 90%, 99%, 99.9% is estimated in Table 1.

Table 1. Calculation of the final biodegradation time corresponding to the conversion of 90, 99 and 99.9% conversion for the different materials.

	A [a]	B [b]		C [a]	
Conv.	Estimated Time (Days)	Estimated Time (Days)	Estimated Time (Years)	Estimated Time (Days)	Estimated Time (Years)
90%	116	227	0.62	1200	3.3
99%	230	5501	15.1	2380	6.5
99.90%	340	100,300	275	3530	9.7

[a] Estimated from first order kinetics. [b] Estimated from the Wilkinson model.

The calculated data for the biodegradation of 90% of the material, 116 days, achieved further correspondence with the experimental data. It is noted, as said above, that it is impossible to estimate the time required to reach a 100% conversion for purely mathematic limits with this model, since it would be necessary calculate the ln(0). However, a concentration of residual polymer that is as low as desired can be entered in the formula, as long as it is different from 0.

The marked difference in the time taken for different conversions is due to the non-linear form of the model with respect to the concentration of the substrate. That is, when almost all the reagent is converted, the rate of the reaction slows down considerably, imposing longer times on the biodegradation of the residual material.

A model of enzyme biodegradation based on the Michaelis–Menten equation was also used to reprocess the data, applied in its linearized form according to a Lineweaver-Burk plot, where v is the degradation rate and v_{max} its maximum value, K_M is the Michaelis constant and [S] is the substrate concentration.

$$\frac{1}{v} = \frac{1}{v_{max}} + \frac{K_M}{v_{max}[S]} \tag{6}$$

This model led to a non-linear plot for substrate A, denoting that it is unsuitable to correctly fit the data. The elaboration led to a broad estimation of the degradation time with respect to the first-order model. Details are reported in the ESI, Figure S1, for the interested reader.

The application of integrated kinetic equations of orders zero and two with respect to the substrate was not satisfactory, as reported in the Electronic Supporting Information (ESI), Figures S2 and S3.

3.2. Material B

With an increase in the complexity of the polymer, we expect longer and possibly less complete biodegradation. The curve of biodegradation of formulate B (Figure 1) indicates an initially faster conversion, which then slows down with respect to sample A.

The same approach was used for data elaboration. In this case, the initial rate of biodegradation was very high, but after approximately 25 days, it underwent an abrupt slowdown, until it became negligible. The application of the same criteria of data-reprocessing throughout the whole experimental field has, however, produced a curve instead of a line, as shown in Figure 3.

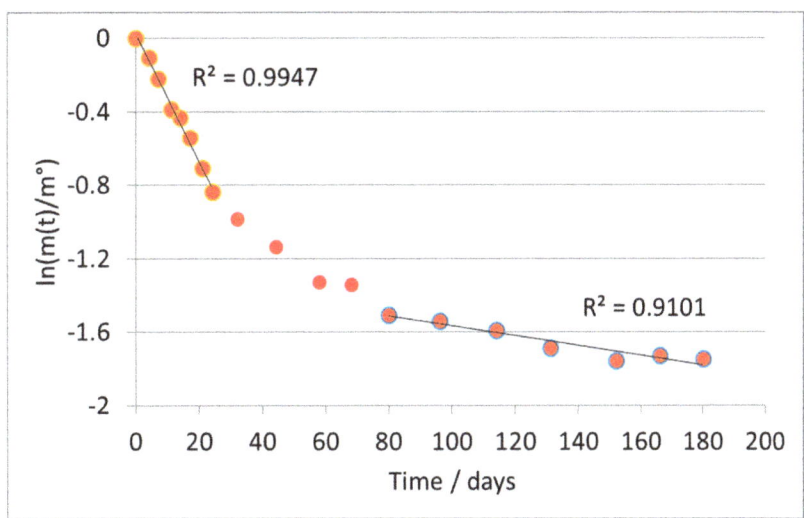

Figure 3. Elaboration of the data of three samples of formulate B according to the first-order model.

This suggests the inadequacy of the first-order kinetic model in this case. Furthermore, the application of equations of order 0 or 2 did not produce significant improvements (see Supplementary Information).

Therefore, we adopted a different strategy. We assumed a reaction of order n

$$dn/dt = k\, n^n \tag{7}$$

and we applied the method of Wilkinson for the determination of kinetic parameters [22].

Approximately, it is possible to calculate a parameter $p = 1-m/m°$ and to perform a linear regression of t/p vs. t, as follows

$$\frac{t}{p} = \frac{n\,t}{2} + \frac{1}{k\,m^{°n-1}} \tag{8}$$

Therefore, the double of the slope of the line provides the estimated reaction order, while it is possible to estimate k from the intercept. According to this approach, we elaborated the kinetics of degradation of material B.

From the values of slope and intercepts, we estimated (Table 2) the values of n and k in the two curve sections (roughly active before and after 30 days (Figure 4).

Table 2. Kinetic parameters for the Wilkinson Model used to interpret the biodegradation of formulate B.

	n	k (day^{-1} mg$^{-0.3}$)
<30 days	0.71	0.14 ± 0.05 (k1)
>30 days	2.5	$(5.37 \pm 0.15)\,10^{-5}$ (k2)

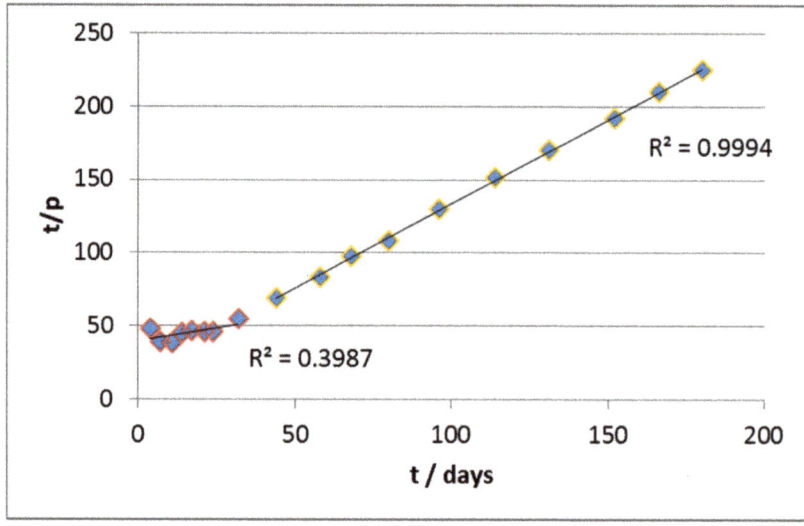

Figure 4. Elaboration of the kinetic data of formulate B according to the Wilkinson model.

In particular, it can be seen that in the early stages of the biodegradation, the estimated reaction order is <1 (as if there were product inhibition). The kinetics are also quite fast thanks to the relatively high value of k (as compared, for example, to material A) and the abundance of the substrate, which mitigates the slowing effect due to a fractional order of reaction. Notice that, in the first segment, the R^2 value of the regression is severely low, partly for the scattered values, but also due to the very small slope of the curve, which is the main reason for the very low value of the correlation coefficient.

In the second section, on the other hand, the apparent order of reaction is higher and this, combined with the progressive consumption of substrate, leads to a significant slowing down of the reaction rate. In parallel, the fact that the curves of biodegradation practically reach a plateau is reflected in a kinetic constant which is orders of magnitude lower with respect to the previous section.

Using these kinetic data, we made two predictions, including the conversion in the experimental field to test the reliability of the model (Table 3):

Table 3. Calculation of the biodegradation time (days) corresponding to the conversion of 20 and 70% of the material in comparison with the experimental datum.

	1st Order Model		Wilkinson Model		
Conversion	Estimated Time Using k1	Estimated Time Using k2	Estimated Time Using k1	Estimated Time Using k2	Experimental
20%	6	-	8.3	-	9
70%	-	83	-	51	58
99%	-	1706	-	5501	-

From these data, it is evident that the first section (t < 30 days) is effectively described by the model. In the second section, the prevision is reasonable, although underestimated by 8–10%. The kinetic data of the second segment were used to estimate the time of biodegradation to 90%, 99% and 99.9% with the results reported in Table 1. By contrast, the first-order model underestimates the biodegradation time to achieve 20% conversion and poorly fits the prediction of 70% conversion.

We are interested in the prediction of the ultimate biodegradation, so the most interesting part of the curve is the one at higher times. The first part is also the most dependent on the possible presence of additives and pro-oxidants. The data reported in Table 3 show that the predicted time for 20% biodegradation is better calculated by the Wilkinson model, even if the almost horizontal fitting curve has a lower correlation coefficient.

Therefore, it is highly recommended to compare the kinetic models, not only for their fitting ability, but for their overall capacity to correctly represent the whole dataset.

Finally, we applied the Michaelis–Menten, zero-order and second-order models also to material B (see ESI), Figures S4–S6. The reprocessing of the data led to the same problems highlighted for material A.

3.3. Material C

The set of data for the formulate C (Figure 1) shows only a partial conversion of the polymer at the end of the test and a sharp downturn in reaction rate after the first 20–25 days. There is a rapid digestion of approximately 20% of the initial mass of the polymer and a subsequent slowdown. This could be due to a sequential mechanism of attack, according to which the chain is first broken and is partially eliminated in a more degradable portion, e.g., an aliphatic chain; subsequently, it proceeds with the biodegradation of the less degradable residual, for instance, an aromatic ring, which is considerably more resistant. This may explain the evolution of the curve, which slows down after 30 days, without reaching a net plateau.

Applying the first-order model (Figure 5) led again to a broken line, and in this case we separately treated the two branches of the curve.

In particular, we discuss the data of the second branch, since the first section does not clearly represent the rate-determining step of the reaction, as already discussed for previous types of material. The average value of the kinetic constant (second branch) was $k = 0.00200 \pm 0.00012$ days^{-1}.

The calculated halving time using k2 was 345 days, but there are no experimental data to check this value. However, this forecast seems to be reasonable, observing the available experimental data. In fact, the biodegradation of ca. 20–25% of the polymer occurs in ca. 155 days (excluding the first period of 25 days, in which the kinetic was much faster and converted ca. 20% of the substrate). It therefore seems likely that 345 days following the slowest kinetics can lead to 50% conversion.

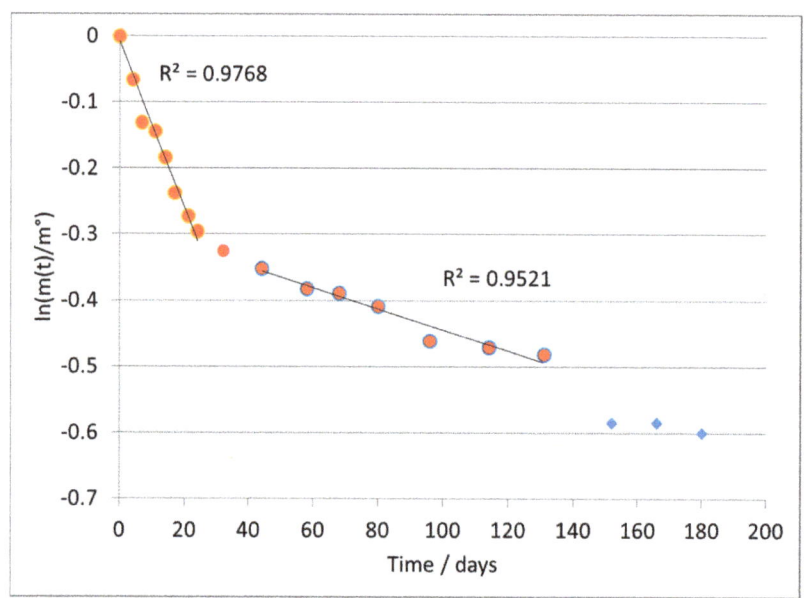

Figure 5. Linear regression of the curve of conversion of material C according to the first-order model. Blue diamonds were considered as outliers and not included in the regression.

Accordingly, we estimated the following conversion times for biodegradation of 90%, 99% and 99.9% of the substrate (Table 1).

We also applied the Wilkinson model to this substrate (Figure 6). The reprocessing led to an apparent value of reaction order n = 4.2, with values of k in the order of 10^{-8} day^{-1} mg^{-3}. Using these data to determine the time taken to reach 20% conversion returns an estimate of 25 days, broadly comparable with the experimental value. However, when one tries to provide the trend in the longer term, the deviations between calculated and experimental data become unacceptable. Even when excluding the last three points, which were again defined as outliers according to statistical analysis of the error distribution, the estimated parameters were n = 5.1 and k in the order of 10^{-10} day^{-1} mg^{-3}, leading to overall unreliable predictions. The calculated time for biodegradation of 20% of the substrate was 20.1 days vs. 21 in the experimental, while for 30% biodegradation it was 44.7 days vs. 80 experimental.

This approach was, therefore, abandoned for material C, in spite of the quite good quality of data regression, also testified by the higher R^2 value with respect to the first-order model, because it does not allow the weighting of the different branches of the curve, which likely rely on different mechanisms.

The Michaelis–Menten, zero-order and second-order models were also unsatisfactory when applied to this sample (see ESI), Figures S7–S9.

3.4. General Remarks

This work does not aim to provide evidence for or to discuss the biodegradability behavior of the tested materials, but rather to propose a kinetic approach to the elaboration of the degradability data (collected on whichever material) for a better comparison between different samples and tests.

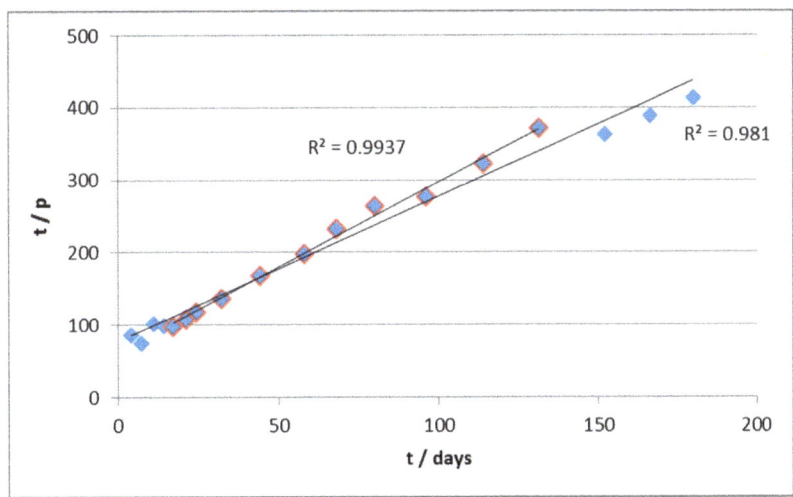

Figure 6. Regression of the curve of conversion of material C according to the Wilkinson model. Last three points included or not included in the regression.

Polyolefins represent the most widely used polymers, especially in the commercialization of single-use manufacts, such as food and beverage packaging. Pure polyethylene (i.e., not treated or combined with pro-oxidants) is extremely recalcitrant to degradation by abiotic or biotic factors, demonstrating limited mass loss after burial in humid soil for several years. Its resistance is due to concomitant factors. The hydrophobic chain prevents attack and colonization by microbials, which also cannot digest high-molecular-weight entities. Indeed, these molecules cannot enter microbial cells to be digested by intracellular enzymes and they are inaccessible to extracellular enzymes because of their excellent barrier properties. This is even more severe when increasing the crystallinity degree. As an example, UV-irradiation of polyethylene for 16 days before burial in soil produced less than 0.5 wt% carbon after 10 years, while in that without any pretreatment, less than 0.2 wt% CO_2 was observed [23].

Thus, in order to circumvent this exceptional stability that brings environmental accumulation, different strategies have been formulated to improve the biodegradability of plastic-based objects prepared from conventional polymers, hence maintaining their functionality and processing advantages.

Possible pretreatments and the use of additives deeply modify the biodegradation of the material, in order to improve the abiotic steps of oxidation and to favour subsequent biotic action. For instance, a high-density PE film was chemically treated by immersion into $KMnO_4$/HCl at for 8 h and 10% citric acid for 8 h at 45 °C. The HDPE degradation gradually increased from 9.4 to 20.8%, indicating synergism between biotic and abiotic factors [5]. Various transition-metal-based additives (traces of Mn, Fe, Co, Ti, e.g., as stearate or oleate) are also well known to improve the biodegradability thanks to an increase in the hydrofilicity and preliminary oxidation of the chain. A total of 70% degradation of a pre-oxidized high-molecular-weight polyethylene sample has been reported after 15 days of treatment with P. chrysosporium strain MTCC-787 [24]. Furthermore, the TDPA technology combines a transition metal carboxylate and an aliphatic poly (hydroxyl–carboxyl acid) [25]: samples of low-density PE films were tested with ca. 50–60% degradation after 18 months in soil. After 70 weeks, no fragments were collected from the soil [26]. Moreover, oxygenated compost columns at 50 °C to degrade PE films after fragmentation returned 60% mineralization after 400 days, whereas thermoformed PP films at 60 °C mineralized up to 60% in 700 days. The difference in degradation times was attributed to the different film thicknesses, since another important parameter is size [27].

It is clear that the search for a solution to the contingent issue of the scarce or nil biodegradability of packaging polymers has led to important results, but a plethora of pro-degradants are commercially available, as recently listed [14], and may be found in commercial formulates [5,28], often without clear indication: this further jeopardises the possibility of predicting the ultimate biodegradation time of such formulates.

In addition, the interest in the development of biodegradable plastics is rapidly raising, with a lot of composite materials that increase the degradability of the manufacts relying on more intrinsically degradable polymers where feasible (e.g., PLA), or on blending classic polyolefins with a biodegradable material (e.g., starch), often in concomitance with pro-oxidants. This wide variability of formulations is often unrecognizable in the disposed manufact, which reports only general indications on the base material (polyethylene, polypropylene, polystyrene, etc.) and on disposal prescriptions. Therefore, a more systematic approach towards degradation testing may help in comparing different formulations. We here propose a simplified kinetic modelling that may be useful to estimate the ultimate degradation time through modelling the aerobic biodegradation tests collected through a standardized method, whatever the tested material. This approach can be used to evaluate every batch of data of CO_2 evolution vs. time.

Very wide formulations are currently commercialised. The presence of degradation enhancers may be one of the possible causes of the different degradation rates in the obtained curves. However, this does not affect the modelling methods, since promoters can have different effects. They can facilitate colonisation and attack by microorganisms or render the surface more prone to degradation (e.g., increasing hydrophilicity). In this case, one should see an overall increase in rate, but the mechanism remains practically the same and a pattern similar to material A is expected. In case the promoter instead implies limited durability (e.g., it degrades itself or gets lost in time, e.g., by leaching), one can expect an enhanced rate at the beginning of testing for the whole life of the material, so the overall curve manifests different slopes in time (as for materials B and C); in such a case, the expected ultimate degradation time should be considered from the slowest portion of the curve. The same variation in the curve is expected when the material is constituted of portions of the chain that can be more promptly attacked and portions that are harsher to degrade.

The particle size also deeply affects the degradation rate; examples of the testing of materials with different grain sizes can be found in the literature [11,13]. Fine comminution exposes more surface area and leads to easier formation of the biofilm, which is ultimately responsible for the biodegradation. Thus, the very small particles used in the present experiments cannot be compared with the rates observed for films, layers, etc. The wide variability of conditions and formulations contributes to the miscellaneous results landscape, making it really difficult to draw a general conclusion on the biodegradability of a material. Therefore, besides using standardized testing protocols, the use of well-defined data elaboration models would help obtain more uniform and comparable results.

The present data were elaborated using various kinetic models, highlighting their strengths and limitations. In every case, the elaboration was based on the best available kinetic models in the state of the art. The choice of the most appropriate kinetic models was made not only on the basis of the examination of the results, but also in light of some indications found in the literature. Specifically, A. Models et al. [29] underline that the biodegradation of polymeric materials in a homogeneous phase is properly describable by means of a Michaelis–Menten type kinetic model (enzyme kinetics), which is not entirely appropriate or properly extendable to the case of the aerobic final biodegradation of plastic materials under controlled composting conditions (see ESI). The application of a kinetic model of the first order represented a better representation of the results.

In support of this, another investigation [30] adopted a first-order kinetics approach. The authors also consider biodegradation that is possible via reactions in series, always of first order, when a period of initial induction is present in the formation of CO_2. In other studies, a purely empirical strategy was adopted, by interpolating the data of CO_2

formation as a function of time with mathematical models (not mechanistic) [31]. In such a case, it would be possible to calculate the time required for the formation of the theoretical amount of CO_2 corresponding to complete biodegradation of the polymer. It should be recalled, however, that the models obtained by pure fitting of experimental data, not based on a verified mechanism of reaction, are usable for the forecasts within the adopted experimental field and not for extrapolation to a final degradation time longer than the experiment. Therefore, they are not the best option for long-term forecasting.

4. Conclusions

The ISO 14855 procedure represents a standard and reliable method to collect kinetic data for the biodegradation of different sets of polyolefin samples and of plastic materials in general. The main limit of this standard procedure is whether CO_2 evolution is fully representative of biodegradation. Due to the fact that part of the carbon remains in the soil as part of the microorganisms, 100% conversion cannot physiologically be obtained. This effect can be managed by replicating the experiments and comparing the results with a blank (same conditions and environment but without the sample) and a reference degradable material.

Established kinetic models to interpret such data are currently scarce in the literature and a more quantitative assessment of the biodegradation pattern may be helpful regarding this point. Indeed, quantification of the rate of decomposition and its evolution over time better allows a comparison of materials. To this end, we proposed a comparison between rather simple kinetic models, to compute relevant kinetic parameters to interpret the biodegradation patterns of different materials. The data, expressed as CO_2 formation as a function of time, can be easily elaborated through first-order kinetics, which were fully suitable to interpret the simplest cases, such as formulate A, which leads to a monotonous degradation curve. Zeroth- and second-order kinetic equations, however, were unsuitable for all the materials.

In some tests, a plateau conversion was reached, with apparent slowdown of the reaction. This can be due to either the exhaustion of compost activity, or to materials' formulation (e.g., the presence of some unknown biodegradation-aiding additives, consumed during the reaction, or the mineralization of the weaker part of the polymer structure, leaving harsher fractions to be degraded slower).

The elaboration of the kinetic data for formulates B and C through a first-order kinetic model was not fully satisfactory, leading to a broken line. The kinetic constant derived from the regression allowed the calculation of reliable estimates of the time of half conversion of the substrate, and the time of final biodegradation of the compound, in every case. The Wilkinson model allowed for interpretation of the more complex curves in the case of materials B and C. The linear regression was particularly important in the second part of the curve, at a longer time, where data relevant to the final biodegradation can be derived. The Wilkinson model suggests the order of magnitude of the apparent reaction order and explains the slowdown of the process. The predictions by this model were more reliable for material B than C, which returned more satisfactory control data through the first-order model. Indeed, for material C, the prediction of the biodegradation time by the first-order model was better at adapting to the experimental data than the value predicted through the Wilkinson model. Finally, the Michaelis–Menten approach was adapted to the present case to estimate the maximum biodegradation rate and was marginally reliable only for the simpler substrates.

Overall, the easy-to-handle elaboration of data collected through a standard test proposed here can help to assess quantitative parameters to estimate the biodegradability of different substrates and to quantitatively compare the materials. It is highly recommended to rely on the whole analysis of the quality of fitting and of predictions, instead of considering strictly statistical parameters. For instance, models characterized by lower correlation coefficients were better able to represent the biodegradability than apparently better-fitting regressions.

As a general procedure, it is suggested to preliminarily interpret the degradation kinetic data through a simple first-order model. If correctly linearized (except after reaching a plateau or during an induction time), the kinetic parameters allow to predict the expected biodegradation time for the conversion of different amounts of material, until ultimate degradation. Instead, if the data give rise to a monotonous curve plot, the Wilkinson model can be used to assess the apparent order of reaction and the kinetic constant. A third option may arise when different degradation mechanisms are active, for instance when a portion of the polymer degrades faster, leaving a harsher residue to decompose. In such a case, the first faster decay rate can be discarded, and the later part of the curve interpreted to assess the ultimate degradation time, as in the example of material C.

Supplementary Materials: The following are available online at https://www.mdpi.com/2673-4117/2/1/5/s1, Decription of Michelis Menten, Zeroth and second order kinetic models, Figure S1: Michelis Menten elaboration according to a Lineweaver Burk plot of the biodegradation data of material A. Red void symbols represent a zoom of the linearization used, Figure S2: Zero order kinetic equation for the biodegradation data of material A, Figure S3: Second order kinetic equation for the biodegradation data of material A, Figure S4: Michelis Menten elaboration according to a Lineweaver Burk plot of the biodegradation data of material B. Red void symbols represent a zoom of the linearization used, Figure S5: Zero order kinetic equation for the biodegradation data of material B, Figure S6: Second order kinetic equation for the biodegradation data of material B, Figure S7: Michelis Menten elaboration according to a Lineweaver Burk plot of the biodegradation data of material C. Red void symbols represent a zoom of the linearization used, Figure S8: Zero order kinetic equation for the biodegradation data of material C, Figure S9: Second order kinetic equation for the biodegradation data of material C.

Author Contributions: Conceptualization, I.R.; methodology, F.C.; data curation, F.C.; writing—original draft preparation, G.R.; writing—review and editing, I.R.; supervision, G.R.; funding acquisition, I.R. All authors have read and agreed to the published version of the manuscript.

Funding: This research received no external funding.

Institutional Review Board Statement: Not applicable.

Informed Consent Statement: Not applicable.

Data Availability Statement: The data presented in this study are available in this paper and its Supplementary information file.

Conflicts of Interest: The authors declare no conflict of interest.

References

1. Sam, S.T.; Nuradibah, M.; Ismail, A.H.; Noriman, N.Z.; Ragunathan, S. Recent advances in polyolefins/natural Polymer blends used for packaging application. *Polym. Plast. Technol. Eng.* **2014**, *53*, 631–644. [CrossRef]
2. Galgani, L.; Loiselle, S. Plastic Accumulation in the Sea Surface Microlayer: An Experiment-Based Perspective for Future Studies. *Geosciences* **2019**, *9*, 66. [CrossRef]
3. Ubeda, B.; Gálvez, J.Á.; Irigoien, X.; Duarte, C.M. Plastic accumulation in the Med sea water. *PLoS ONE* **2015**, *10*, e0121762.
4. Barnes, D.K.A.; Galgani, F.; Thompson, R.C.; Barlaz, M. Accumulation and fragmentation of plastic debris in global environments. *Philos. Trans. R. Soc. B Biol. Sci.* **2009**, *364*, 1985–1998. [CrossRef]
5. Ghatge, S.; Yang, Y.; Ahn, J.H.; Hur, H.G. Biodegradation of polyethylene: A brief review. *Appl. Biol. Chem.* **2020**, *63*, 1–14. [CrossRef]
6. Restrepo-Flórez, J.M.; Bassi, A.; Thompson, M.R. Microbial degradation and deterioration of polyethylene—A review. *Int. Biodeterior. Biodegrad.* **2014**, *88*, 83–90. [CrossRef]
7. Barton-Pudlik, J.; Czaja, K.; Grzymek, M.; Lipok, J. Evaluation of wood-polyethylene composites biodegradability caused by filamentous fungi. *Int. Biodeterior. Biodegrad.* **2017**, *118*, 10–18. [CrossRef]
8. Fontanella, S.; Bonhomme, S.; Brusson, J.-M.; Pitteri, S.; Samuel, G.; Pichon, G.; Lacoste, J.; Fromageot, D.; Lemaire, J.; Delort, A.-M. Comparison of biodegradability of various polypropylene films containing pro-oxidant additives based on Mn, Mn/Fe or Co. *Polym. Degrad. Stab.* **2013**, *98*, 875–884. [CrossRef]
9. Fontanella, S.; Bonhomme, S.; Koutny, M.; Husarova, L.; Brusson, J.-M.; Courdavault, J.-P.; Pitteri, S.; Samuel, G.; Pichon, G.; Lemaire, J.; et al. Comparison of the biodegradability of various polyethylene films containing pro-oxidant additives. *Polym. Degrad. Stab.* **2010**, *95*, 1011–1021. [CrossRef]

10. Koutny, M.; Sancelme, M.; Dabin, C.; Pichon, N.; Delort, A.-M.; Lemaire, J. Acquired biodegradability of polyethylenes containing pro-oxidant additives. *Polym. Degrad. Stab.* **2006**, *91*, 1495–1503. [CrossRef]
11. Funabashi, M.; Ninomiya, F.; Kunioka, M. Biodegradability evaluation of polymers by ISO 14855-2. *Int. J. Mol. Sci.* **2009**, *10*, 3635–3654. [CrossRef] [PubMed]
12. Castro-Aguirre, E.; Auras, R.; Selke, S.; Rubino, M.; Marsh, T. Insights on the aerobic biodegradation of polymers by analysis of evolved carbon dioxide in simulated composting conditions. *Polym. Degrad. Stab.* **2017**, *137*, 251–271. [CrossRef]
13. Arutchelvi, J.; Sudhakar, M.; Arkatkar, A.; Doble, M.; Bhaduri, S.; Uppara, P.V. Biodegradation of polyethylene and polypropylene. *Indian J. Biotechnol.* **2008**, *7*, 9–22.
14. Ammala, A.; Bateman, S.; Dean, K.; Petinakis, E.; Sangwan, P.; Wong, S.; Yuan, Q.; Yu, L.; Patrick, C.; Leong, K.H. *An Overview of Degradable and Biodegradable Polyolefins*; Elsevier Ltd.: Amsterdam, The Netherlands, 2011; Volume 36.
15. Dřímal, P.; Hoffmann, J.; Družbík, M. Evaluating the aerobic biodegradability of plastics in soil environments through GC and IR analysis of gaseous phase. *Polym. Test.* **2007**, *26*, 729–741. [CrossRef]
16. Nguyen, D.M.; Do, T.V.V.; Grillet, A.-C.; Ha Thuc, H.; Ha Thuc, C.N. Biodegradability of polymer film based on low density polyethylene and cassava starch. *Int. Biodeterior. Biodegrad.* **2016**, *115*, 257–265. [CrossRef]
17. Ali, H.E.; Abdel Ghaffar, A.M. Preparation and Effect of Gamma Radiation on The Properties and Biodegradability of Poly(Styrene/Starch) Blends. *Radiat. Phys. Chem.* **2017**, *130*, 411–420. [CrossRef]
18. Kawai, F.; Watanabe, M.; Shibata, M.; Yokoyama, S.; Sudate, Y. Experimental analysis and numerical simulation for biodegradability of polyethylene. *Polym. Degrad. Stab.* **2002**, *76*, 129–135. [CrossRef]
19. Ramis, X.; Cadenato, A.; Salla, J.M.; Morancho, J.M.; Vallés, A.; Contat, L.; Ribes, A. Thermal degradation of polypropylene/starch-based materials with enhanced biodegradability. *Polym. Degrad. Stab.* **2004**, *86*, 483–491. [CrossRef]
20. Kunioka, M.; Ninomiya, F.; Funabashi, M. Novel evaluation method of biodegradabilities for oil-based polycaprolactone by naturally occurring radiocarbon-14 concentration using accelerator mass spectrometry based on ISO 14855-2 in controlled compost. *Polym. Degrad. Stab.* **2007**, *92*, 1279–1288. [CrossRef]
21. Massardier-Nageotte, V.; Pestre, C.; Cruard-Pradet, T.; Bayard, R. Aerobic and anaerobic biodegradability of polymer films and physico-chemical characterization. *Polym. Degrad. Stab.* **2006**, *91*, 620–627. [CrossRef]
22. Connors, K.A. *Chemical Kinetics*; VCH Publishers, Inc.: Hoboken, NJ, USA, 1990.
23. Albertsson, A.; Karlsson, S. The influence of biotic and abiotic environments on the degradation of polyethylene. *Prog. Polym. Sci.* **1990**, *15*, 177–192. [CrossRef]
24. Mukherjee, S.; Kundu, P. Alkaline fungal degradation of oxidized polyethylene in black liquor: Studies on the effect of lignin peroxidases and manganese peroxidases. *J. Appl. Polym. Sci.* **2014**, *131*, 40738. [CrossRef]
25. Garcia, R.; Gho, J. Degradable/Compostable Concentrates, Process for Making Degradable/Compostable Packaging Materials and the Products Thereof. U.S. Patent 5,854,304A, 29 December 1998.
26. Chiellini, E.; Corti, A.; Swift, G. Biodegradation of thermally-oxidized, fragmented low-density polyethylenes. *Polym. Degrad. Stab.* **2003**, *81*, 341–351. [CrossRef]
27. Available online: https://www.reverteplastics.com/ (accessed on 1 February 2021).
28. Scott, G.; Wiles, D.M. Programmed-life plastics from polyolefins: A new look at sustainability. *Biomacromolecules* **2001**, *2*, 615–622. [CrossRef] [PubMed]
29. Modelli, A.; Calcagno, B.; Scandola, M. Kinetics of Aerobic Polymer Degradation in Soil by Means of the ASTM D 5988-96 Standard Method. *J. Envrion. Polym. Degrad.* **1999**, *7*, 109. [CrossRef]
30. Leejarkpai, T.; Suwanmanee, U.S.; Rudeekit, Y.; Mungcharoen, T. Biodegradable kinetics of plastics under controlled composting conditions. *Waste Manag.* **2011**, *31*, 1153. [CrossRef]
31. Mohee, R.; Unmar, G.D.; Mudhoo, A.; Khadoo, P. Biodegradability of biodegradable/degradable plastic materials under aerobic and anaerobic conditions. *Waste Manag.* **2008**, *28*, 1624–1629. [CrossRef]

Article

A Critical Review of the Equivalent Stoichiometric Cloud Model Q9 in Gas Explosion Modelling

Vincent H. Y. Tam [1,*], Felicia Tan [2] and Chris Savvides [†]

[1] School of Engineering, Warwick University, Coventry CV4 7AL, UK
[2] BP, I&E Engineering, London SW1Y 4PD, UK; felicia.Tan@uk.bp.com
* Correspondence: v.tam@warwick.ac.uk
[†] Retired, formerly BP, no email address.

Abstract: Q9 is widely used in industries handling flammable fluids and is central to explosion risk assessment (ERA). Q9 transforms complex flammable clouds from pressurised releases to simple cuboids with uniform stoichiometric concentration, drastically reducing the time and resources needed by ERAs. Q9 is commonly believed in the industry to be conservative but two studies on Q9 gave conflicting conclusions. This efficacy issue is important as impacts of Q9 have real life consequences, such as inadequate engineering design and risk management, risk underestimation, etc. This paper reviews published data and described additional assessment on Q9 using the large-scale experimental dataset from Blast and Fire for Topside Structure joint industry (BFTSS) Phase 3B project which was designed to address this type of scenario. The results in this paper showed that Q9 systematically underpredicts this dataset. Following recognised model evaluation protocol would have avoided confusion and misinterpretation in previous studies. It is recommended that the modelling concept of Equivalent Stoichiometric Cloud behind Q9 should be put on a sound scientific footing. Meanwhile, Q9 should be used with caution; users should take full account of its bias and variance.

Keywords: gas explosion; equivalent stoichiometric cloud; Q9; explosion risk assessment; model evaluation

1. Introduction

Gas explosion risk assessment forms a key part of major hazard risk assessments in the oil and gas and petrochemical industries. Central to this assessment is the quantification of consequences of gas explosions, such as loading (i.e., drag and overpressure) impact on structures, equipment and buildings. This assessment includes the calculation of consequences of chains of events prior to gas explosions (examples include failures of containment, formation of flammable gas clouds, their ignitions), as well as those proceeding them (for example: overpressures, blasts, wind, and their impact on equipment, structures, etc.). Each of these steps is complicated to model mathematically, many of them often requiring computational fluid dynamics tools.

Simplifications are often made in order to make the assessment tractable within time and resource constraints. This is because a typical explosion risk assessment, say for an offshore facility, could assess thousands of scenarios.

This paper describes one of the simplifications commonly deployed in probabilistic explosion risk assessment (ERA): the representation of a flammable gas cloud for an explosion calculation. A flammable gas cloud could from a boiling or evaporating pool of liquid or a pressurised release (e.g., flange leak from a pressurised system). In terms of frequency of occurrence, pressurised gas release is most common.

The characteristic relevant to ERA of a flammable gas cloud is complex: the cloud is non-uniform in shape and embedded within it are variable gas concentrations and turbulence (distribution and intensity in space and time).

The simplification involves the transformation of a flammable gas cloud with complex shape and concentration distribution into a regular cuboid shaped cloud with uniform concentration typically at or slightly above stoichiometry in a quiescent state (see Figure 1).

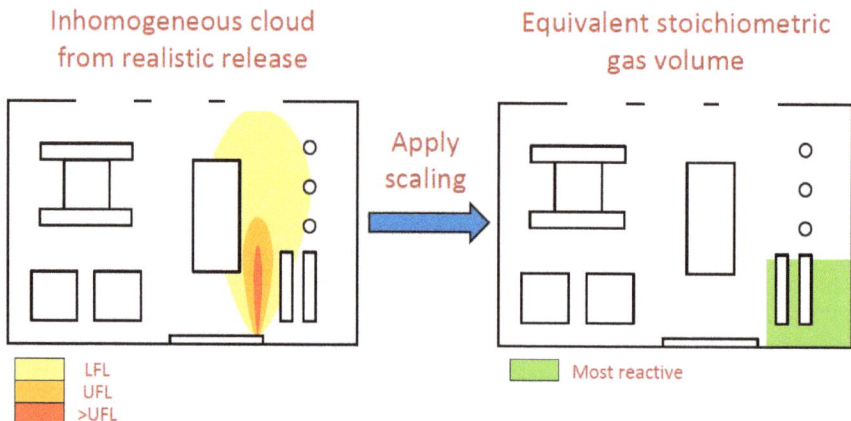

Figure 1. The representation of a complex flammable gas cloud with a simple gas cloud with uniform concentration in a quiescent state (courtesy of J Stewart, Health and Safety Laboratory).

This simplified flammable gas cloud is called the Equivalent Stoichiometric Cloud (ESC). The application of the ESC concept is widespread and enshrined in standards such as NORSOK, in Norway [1], but the definition of the ESC is left open. As the use of computational fluid dynamics (CFD) codes proliferates, more complex representation of ESC develops with time.

Here, we will refer to different ESC representations as ESC models. They are often simple, consisting of one or a few simple equations. They allow the rapid calculation of ESC volumes.

1.1. One specific ESC Model—Q9

This paper focuses on one ESC model commonly referred to as Q9, developed by GexCon for the CFD gas explosion simulator FLACS (FLACS is a computer package. FLACS stands for FLame Acceleration Simulator.) This is because FLACS is overwhelmingly the dominant commercial CFD explosion code used by high hazard industries globally, e.g., the oil and gas, petrochemical, mining industries. Q9 is recommended by GexCon and described in the FLACS manual. As a result, Q9 has become the de facto standard in hazard and risk analyses by the vast majority of consultants in the world.

1.2. Importance of ESC

A clear understanding of this particular ESC model is important as any systematic biases introduced by the Q9 (or any ESC) model show up in accumulated risks, affecting shapes of risk exceedance curves, impacting on engineering designs, emergency response and process safety management. These will be discussed later.

1.3. Objective

The objectives of this paper are two-fold.

(a) Provide a detailed evaluation of Q9 with experimental data from the large-scale experimental data set (BFTSS Phase 3B) (More details are given in Section 5.1) which specifically addressed scenarios that Q9 is designed for.
(b) To clear up confusion generated by two publications: one (by authors of this paper) which concludes that Q9 under-predicts systematically [2] and the other which

concludes that Q9 is overconservative (i.e., over-predicts) [3]. The latter paper did not cite the former, hence creating the current unchallenged impression that Q9 is over-conservative for readers who are not aware of the first.

There could be legitimate reasons for this difference. They will be addressed later in this paper.

1.4. Structure of This Paper

Before going into details, we note that there are many occurrences of abbreviations for frequently used phrases and terms in this paper. Though full definitions are given at points where abbreviations are first used, abbreviations and their definitions are given in the Abbreviation part (Section 10) for the convenience of readers.

This paper is organised in three main parts. The first describes the historic development of the ESC concept, the underlying assumptions behind ESC and the reason for examining a specific ESC model, Q9.

The second part describes the methodology used in our study and a summary of results.

This is followed by a discussion which includes interpretation of results, learning from this study and finally recommendations for taking this subject further.

We will begin by briefly going over the history of ESC.

2. Evolution of Equivalent Stoichiometric Cloud

The simplification of a real flammable gas cloud is a pragmatic approach to a complex problem. This approach is as old as explosion modelling in safety analysis.

2.1. An Historic Perspective

Following a major explosion accident in a chemical factory near Flixborough in the UK in 1974, a version of an equivalent explosion cloud was developed; this is in the form of the TNT equivalent method. This was to assist in the enquiry of the accident [4] and for the assessment of explosion hazards in onshore facilities. This method assumes that the size of a flammable vapour cloud involved in an explosion is represented by the chemical energy of the total mass of flammable material released. This was later refined to the mass of flashed vapour [5,6].

This marked the beginning of the application of the concept of ESC, though the term ESC was coined later.

The TNT equivalent method was later further refined to the volume or mass of cloud within the congested volume in the multi-energy model developed by the TNO Prins Maurits Laboratory in the Netherland [7]. As knowledge improved, the simple multi-energy model evolved to more complex methods, such as GAMES [8] which incorporates effects of equipment density and layout into the multi-energy model. There are other models developed along similar line to the multi-energy concept. It is beyond this scope of this paper to name them all.

This approach continues to be applied to date to onshore facilities for the assessment of consequence and offsite risks at a distance from the location of the explosion.

A different type of explosion model is required for offshore facilities. Owing to limited footprint and close proximity of gas explosion hazards to equipment, protective structures and people, the assessment of explosion loading within and very close to the exploding cloud is needed for assessment of impact on them. Phenomenological and CFD models were developed for these "near-field" applications. With advances in computers, CFD explosion models are widely used. With time, they are used to simulate progressively more and more complicated scenarios.

The underlying development path of ESC models for offshore mirrored that for onshore, namely, to refine the simple representation of a flammable cloud, progressively removing perceived conservatism with time.

Prior to the large-scale gas explosion JIP (JIP—Joint Industry Project called the Blast and Fire for Topside Structure Phase 2 JIP (BFTSS Phase 2)) in the 1990s, a typical assessment of explosion loading would have included a range of sizes of cuboids, the gas cloud containing uniform mixtures of stoichiometric flammable gas and air. An example of this is in the Piper Alpha enquiry conducted by Lord Cullen [9] in which the Christian Michelsen Institute (CMI) submitted explosion loading results from FLACS simulations based on a number of uniformly mixed stoichiometric cuboid volume clouds. The largest cuboid filled the entire volume of the module on the platform. This 100% area filled scenario is called the theoretical worst-case scenario.

If the inventory was not sufficient to fill the entire area, maximum cloud sizes were typically determined by volumes of flammable inventories within isolatable sections and assuming these inventories formed stoichiometric flammable mixtures with air. These smaller cloud sizes, shaped in cuboids, are called specific theoretical worst cases; an example of this application is shown in the design of the Andrew platform in the North Sea [10].

To distinguish this approach from later ones, we will call this "inventory-based ESC volumes".

The results of BFTSS JIP Phase 2 showed that all gas explosion models grossly underpredicted experimental results, some by two or three orders of magnitude [11]. This was the case even for advanced CFD models (including FLACS) which incorporated the state-of-the-art representation of the underlying physics at the time. The theoretical and virtually all the specific worst cases would produce explosion overpressure many times higher than previously estimated, some much higher than the capacities of structures of installed facilities and those being designed at the time.

2.2. Evolution and Type of ESC Volume Methodology

The objective of BFTSS Phase 2 was to provide data at a realistically large scale to validate gas explosion models for application to offshore facilities. Additionally, a subsidiary objective was to evaluate explosion models commonly used at the time. This JIP identified gaps in the industry; the items relevant to our discussion here are:

(i) Accuracies of predictive models: Phase 2 spurred further research in both data generation and analysis, e.g., [12,13], and development of predictive models and tools (e.g., a many-year extension to the Gas Safety Programme at the Christian Michelsen Research (GexCon is a spinoff from it)).

(ii) Procedures of assessment: while the high overpressure observed caused concern, there was a common acknowledgement that the theoretical worst-case scenarios used in Phase 2 to test the explosion model is not representative of real-life situations in accidents.

There was a concerted move towards defining an ESC model which better reflects the formation of flammable gas clouds in real accident situations where (the perception was that) significant portions of inventories released do not take part in explosions, or release their energy at a slower rate than at the optimal stoichiometric concentration—"realistic scenarios".

Application of realistic scenarios required the development of a methodology for **dispersion-based ESC volumes** (henceforth simply referred to as ESC volumes).

Unlike the inventory-based ESC volumes, the dispersion-based ESC volumes would need to account for release conditions (e.g., hole sizes, pressure, direction, etc.), environmental conditions (e.g., platform layout, wind velocities, etc.), non-uniform distributions of gas concentration and flow fields characteristics generated by ambient wind and pressurised releases which interact with equipment and platform layout.

Though the process may appear complex, conceptually it is simple. Instead of inventories released, the ESC volumes are calculated using results from dispersion models; this requires an additional calculation step. Dispersion models can range from simple (e.g., the zonal model, workbook [14], etc.) to complex (e.g., CFD).

With a CFD model, it is possible, in principle, to use the results from a CFD dispersion model directly as input. In practice, this is impractical for risk assessment due to the large number of scenarios considered, high resources required and long calculation time.

Hence, the approach adopted is to simplify the complex cloud into a simple representation of it. There are many ways to reduce the complexity of a dispersing flammable gas cloud to a simple uniform cuboid representation.

As there were no data for this type of scenario, research was carried out to gather data of flammable cloud volumes and explosion overpressures from pressurised gas releases. During this period, a number of ESC models were developed.

GexCon developed ESC models based on flame speed and expansion ratio; starting off with the simplest ERFAC, then Q5, Q8 and Q9 (see Figure 2), progressively increasing the effect of the gas concentration and expansion ratio in the model that defines ESC volumes. These names may appear esoteric; they are taken from variable names used in the Flacs engine. Flacs refers to the numerical explosion simulator. The whole modelling package including pre and post processors is called FLACS. The logic behind this is that it is known that the severity of a gas explosion depends on flame speed which varies with gas concentration.

Equivalent Cloud Metric	Equation
ERFAC (kg) [7]	$M_{ERFAC} = \dfrac{1}{S_{max}} \sum_{i=1}^{n} (fuel_{mass} \times S)_i$
Q5 (m³) [7]	$V_{Q5} = \dfrac{1}{S_{max} E_{max}} \sum_{i=1}^{n} (fuel_{vol} \times S \times E)_i$
Q9 (m³) [7]	$V_{Q9} = \dfrac{1}{(SE)_{max}} \sum_{i=1}^{n} (fuel_{vol} \times S \times E)_i$
Q8 (m³) [7]	$V_{Q8} = \dfrac{1}{E_{max}} \sum_{i=1}^{n} (fuel_{vol} \times E)_i$ where $LFL \leq C \leq UFL$
> LFL (m³) [1]	$V_{>LFL} = \sum_{i=1}^{n} (fuel_{vol})_i$ where $C \geq LFL$
ΔFL (m³) [1]	$V_{\Delta FL} = \sum_{i=1}^{n} (fuel_{vol})_i$ where $LFL \leq C \leq UFL$

Figure 2. Definition of various equivalent stoichiometric cloud (ESC) volume models (table taken from [15], please note that the references in the table are different from those in the reference section in this paper. They as follows: [12] is (Tam et al., 2008) and [16] is (Hansen et al., 2013). In the equations above, M is the mass in kg, V is volume in m3, S is the laminar burning velocity, and E is the stoichiometric ratio of the fuel/air mixture. Their suffixes refer to specific quantities, e.g., V_{Q9} is the volume of Q9, S_{max} is the maximum laminar burning velocity of the fuel in air.

There are simpler ESC models, such as ">LFL" (volume bounded by LFL) and "ΔFL" (volume bounded by the upper flammability limit (UFL) and lower flammability limit (LFL)). They were used by BP following their analysis of large-scale experimental data. Definitions of these ESC models are given in Figure 2.

Presently, Q9 is used widely, other models (e.g., "ΔFL") have only a few users. While there are slightly more complicated methods for flammable volumes which can be transformed into ESC volumes (e.g., a workbook approach [14]), they are hardly used.

3. Underlying Assumption of ESC Methodology

There is an unstated assumption in the current use of ESC (i.e., dispersion-based ESC): that a real flammable gas cloud (with complex distribution of concentration and turbulence) is equivalent to a uniformly mixed stoichiometric flammable gas cloud in quiescent state (Figure 1). There is no theoretical basis or conceptual scientific deduction behind this assumption. Put simply, it is purely an assumption created for expediency (untested prior to the Phase 3B JIP [17]). In many engineering applications, this approach is acceptable when this is backed up by experimental data, leading to an empirical relationship that can then be applied more generally.

Hence, the evaluation of the ESC model performance against experimental data is important.

4. Case for Re-Assessment of Q9

Q9 is widely used, and a de facto standard method for CFD particularly FLACS [3,18]. There is a widely held view that Q9 is over-conservative; we frequently see this assertion in reports submitted to us for review. Part of this justification is that Q9 is recommended (i.e., in the FLACS user guide/manual), the other is the scientific basis of the model which includes some of the obvious physical processes involved. However, the evidence supporting this is not clear cut. There are two papers which give opposite conclusions: The Hansen paper published in 2013 [3] supports the current widely held view, while a paper by the authors published in 2008 [2] gives an opposite conclusion. It is interesting to note that the former paper [3] stated that Q9 is conservative upfront but did not provide evidence in the paper to support it.

Both papers derived their results using different methods; this makes direct comparison difficult. Ref. [2] presented only a comparison of overpressures between data and FLACS results based on Q9 cloud volumes derived from experimental gas concentration measurements. Results presented in [3] were based on simulation of non-homogeneous cloud formed from jet releases and did not involve Q9. The assessment of the performance of Q9 in the Hansen paper relied on a "good" correlation between overpressure data and Q9 volumes from concentration data. However, statistics of the correlation were not given. The Q9 simulations that were presented were not based on "realistic scenario" experiments. Thus, there is no direct comparison.

In addition, the results of the two papers were presented in different formats. While [2] presented them in MV (mean-variance) diagrams (described in more detail in Section 5.4) as recommended by the gas explosion model evaluation (MEGGE) project [19]. Hansen et al [3] presented results in the form of comparisons between observed and predicted pressure readings on selected tests. Therefore, it is difficult to compare the results of these two papers. This problem was highlighted by the UK Health and Safety Executive recently [15].

To avoid any confusion, it is necessary to carry out a comprehensive analysis. For consistency, we compared results from these two papers and additional analysis on MV diagrams. An MV diagram is a common format recognised in all model evaluation protocols [19–21] and that used in the BFTSS JIP Project model evaluation exercise [22].

5. Methodology

5.1. Dataset from Phase 3B

A major part of the Phase 3B of the Blast and Fire for Topside Structure joint industry (BFTSS) project (henceforth referred to as Phase 3B) was a large-scale experiment designed to study realistic scenarios involving the release of high-pressure natural gas into a large-scale model of an offshore production module [17]. Phase 3B provided realistic release scenario data for the development and evaluation of gas explosion models. Both papers ([2,3]) used this dataset.

The whole Phase 3B project consisted of laboratory, medium and large-scale tests. Only the large-scale experimental dataset is used in this paper. Figure 3 shows the experimental test rig which measured about 28 m long, 12 m wide and 8 m high. It was a simplified full-scale model of a compression module on a platform operating in the North Sea at the time. Natural gas was released within the module. The release rate was held constant within each test until gas concentrations and their distribution inside the module reached a steady state prior to ignition. Release rates varied between 2.1 kg s^{-1} and 11.7 kg s^{-1} in the Phase 3B programme. Release directions were in line with one of the three orthogonal coordinate axes of the test rig. In total, twenty tests were carried out. Gas concentrations were measured at 50 locations prior to ignition and overpressure at 25 locations distributed inside the module. Earlier phases of the BFTSS JIPs are not relevant for this paper. The earlier phases provided an interim engineering guidance note [23] and data from experimental programmes addressing theoretical worst-case and specific worst-case scenarios (for model evaluation), and mitigation options [12,19].

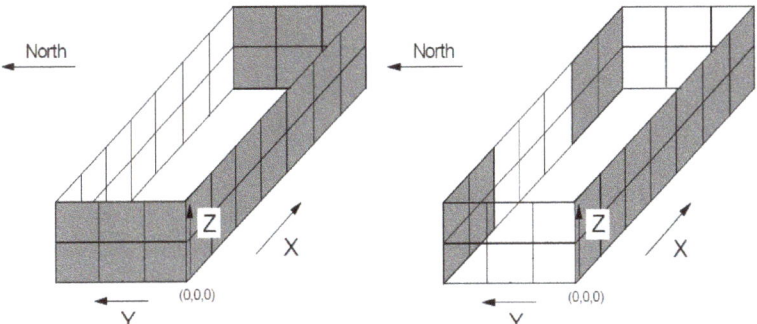

Figure 3. A picture of the Phase 3B test rig. It measures about 28 m long, 12 m wide and 8 m high and was a large-scale model of a process module on an offshore platform. Schematic diagrams showing the two wall layouts or configurations (these layouts are given names "C2" on the left, and "C1" on the right). The direction north points out of the page.

5.2. Previous Work on Flammable Volumes

Just before the Phase 3B JIP, the Dispersion JIP [24] studied the dispersion of releases of pressurised natural gas in the same large-scale module as Phase 3B. ESC models of ">LFL" and ΔFL were evaluated and compared with predictions from FLUENT [25] (and FLACS [26]. These papers showed that both FLUENT and FLACS were able to estimate ">LFL" and ΔFL with little bias. Q9 was not part of the evaluation.

The results of Hansen et al., 2013 indicated that FLACS also could predict Q9 with little bias.

5.3. Versions of FLACS Used

The underlying physics of the explosion code, Flacs, has changed little during the period since the completion of the Phase 3B JIP. FLACS has undergone many development iterations. Significant effort has gone into making the code easier to use, numerically more stable and faster to execute.

Three different sets of results on Q9 had been published in 2001, 2008 and 2013 using FLACS version current at the time. Further data were generated for this paper using the latest version of FLACS.

We compared gas explosion results on selected project work in BP using the current version FLACS 10.6 with FLACS 8.1 which is close to FLACS 99 r2 when GexCon first carried out a comparison of FLACS with Phase 3B data [17].

A point to point comparison gave differences of less than 5% between the "old" and the current version of FLACS. On average, the current version gave slightly lower prediction at pressure above 2 bar and slightly higher below it. These differences are insignificant for our purpose here.

We concluded that results presented in the Phase 3B report [17], Ref. [2] in 2008, and Ref. [3] in 2013 are still valid for comparison purposes. They can be used in conjunction with those generated using the current FLACS code.

5.4. Format of Comparison—The MV Diagram

The results are presented in MV diagrams in this paper. The MV diagram shows two quantities: geometric mean (M_G) and geometric variance (V_G). Their definitions are given in [2] and summarised here:

$$M_G = \exp(<\ln(P/O)>)$$

$$V_G = \exp(<(\ln(P/O))^2>)$$

where:

P = predicted overpressure
O = observed or measured overpressure
<X> denotes expectation value of X

The position on an MV diagram shows the overall performance of mathematical models (in this case ESCs) based on results of comparing large number of simulation—experiment pairs.

An MV diagram gives two important measures: bias and variance. Bias shows whether a model systematically under or over predicts experimental results overall; however, bias does not tell you how likely the results are to over or underpredict data in any one simulation situation; variance provides an indication of this.

Variance shows the range of scatter about the mean. A model with a large bias may appear to be a poor tool; however, a correction factor could be applied in practice if the variance is small. A model which has a large variance is unreliable irrespective of bias; results are difficult to interpret, and no simple correction factors can be applied.

An ideal model will be neutrally biased (with a geometric mean M_G of one) and a variance (geometric variance V_G) of one on the MV diagram. This corresponds to a situation where a model accurately predicts experimental outcome/data with no error every time. Position E shows the position of such a perfect model on the MV diagram (see Figure 4).

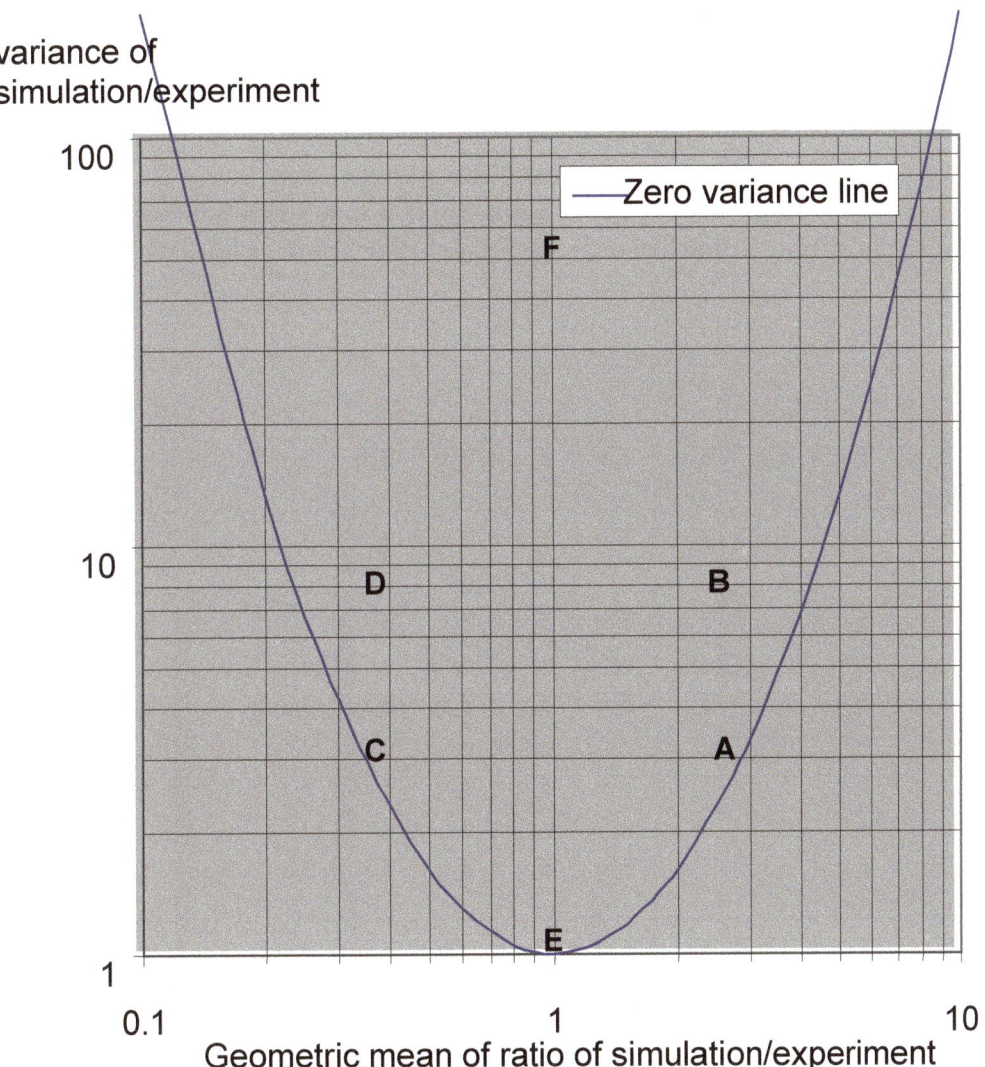

Figure 4. An example MV diagram (taken from [2]). The curve is the zero-variance line. Point A is close to the zero-variance line: it indicates the model consistently overpredicts and has a very low probability of underprediction. Point B is above A and has a high variance: it indicates that the model, though it has a tendency to overpredict, has a wide range of prediction; some are underpredicted—model B is less predictable than Model A. Points C and D are similar to Points A and B, but underpredict. Point E is close to the bottom of the lowest parabola; this indicates consistently accurate prediction of experimental data. Point F is unbiased; it has a very high variance; a model with this property is of little use in practice as it behaves like a random number generator.

Continuing with Figure 4, position C shows the relative position of a model with bias M_G much less than 1 but with a small variance V_G, indicating the model is underpredicting data systematically and consistently. Position F is for one with M_G close to 1 showing that there is no systematic bias, but it has a large V_G. This means that the model in position F produces unreliable results which can wildly under- or over-predict.

There are many reasons MV diagrams are preferred in model evaluation protocols. They are described in detail in [27]. Here are a couple of key considerations:

(a) Scale

As the range of experimental data spans more than three order of magnitudes (from millibars to tens of bars), predicted-over-observed graphs tend to have data points bunching together into groups. The alternative of using log plots compresses deviations making agreement better than superficial appearance suggests.

(b) Ease of comparison of the two key characteristics of model predictions

There are many statistical methods for comparing predicted and observed results, such as quantile–quantile plots, residual values, scatter diagrams, bar charts, etc. While they are useful in detailed examination of model behaviour/characteristics, they contain too much information. Figure 5 is an example which shows model comparison results between predicted and observed. It is difficult to pick up visually relative biases and variances from it.

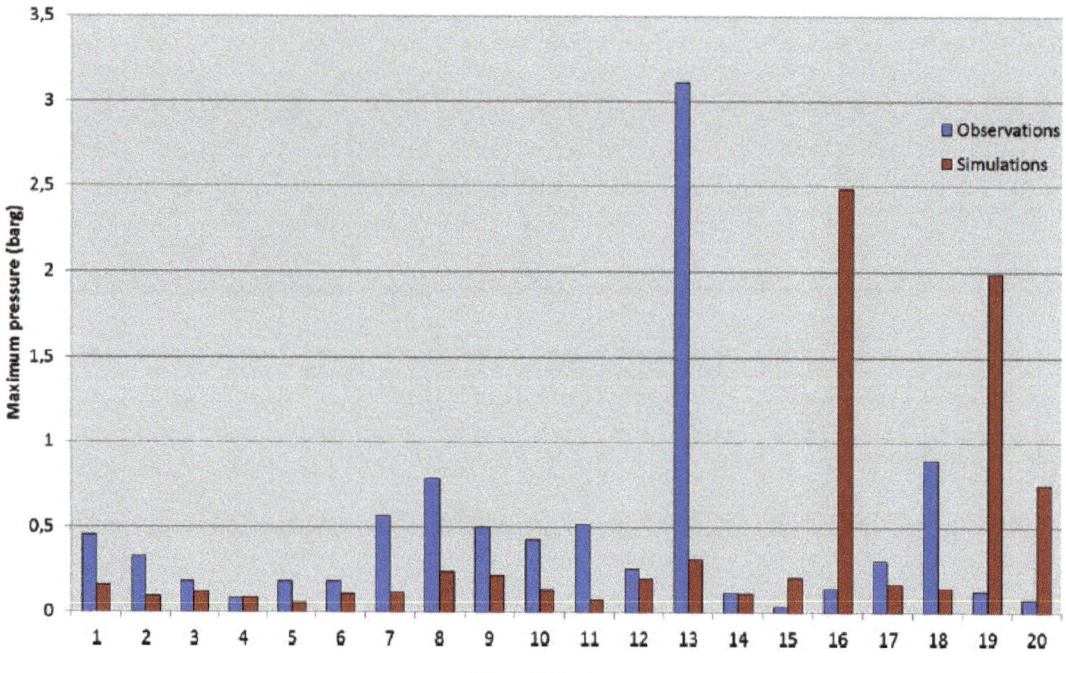

Figure 5. *Cont.*

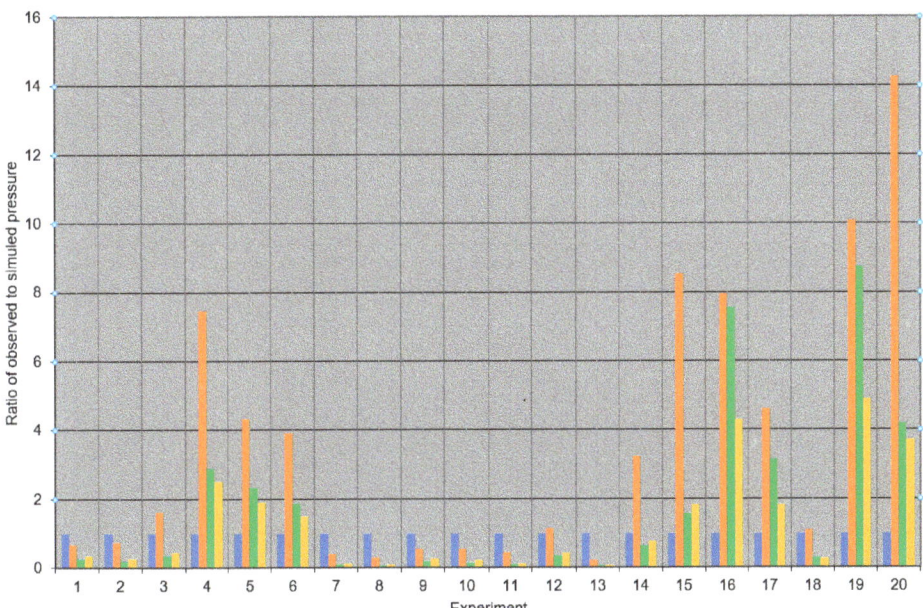

Figure 5. (top): A figure showing a comparison of calculated maximum overpressure with measurement (reproduced from [3], and Figure 5 (**bottom**): the comparison of various ESC models against data (by the author) showing the ratio of model prediction with measured values. The number on the x-axis represents the test number in the experiment. The blue bar in the bottom figure is a reference showing the perfect agreement position of 1. Each colour refers to one ESC model; the exact ESC model is not important here. Both of these figures are confusing for the purpose of comparing model performance. The purpose of showing these two figures is to illustrate the difficulties in extracting overall biases and variance for various models.

6. Results and Evaluation

6.1. Estimation of Overpressures by Direct FLACS Dispersion-Explosion Simulations

One method of calculating explosion overpressure in a realistic scenario is to use the results directly from a dispersion calculation, without involving any ESC model.

A large part of the results presented by [3] was based on this methodology. This involves calculating the evolution of a flammable gas cloud from a known release source (with known orifice diameter and release rates) and release conditions. An electric spark ignites the cloud once it has reached a steady state. This methodology does not require the use of Q9 or any ESC models. There are two sources of data: (a) Figure 5 (top), and (b) the Phase 3B report [16].

A summary of the results is given in Table 1.

Table 1. Geometric Mean and Variance: direct dispersion–explosion simulations.

Source	Hansen et al., 2013 [3]		Phase 3B Report [17]	
	M_G	V_G	M_G	V_G
All	0.73	12.2	0.68	11.7
C1 only	0.38	3.50	0.46	2.73
C2 only	1.17	26.5	1.10	63.6

Based on 0.5 m control volume.

The Hansen 2013 [3] result gives an overall geometric mean (M_G) of 0.73 and a geometric variance (V_G) of 12. Though the underprediction is modest, the variance is large. Dividing the dataset into the two layout configurations C1 and C2 in the large-scale tests (see Figure 3), the corresponding M_G and V_G values for C1 are 0.38 and 3.5 and for C2 are 1.2 and 27. The underprediction is more severe for C1 than that for the complete dataset and with a reduced variance. However, the results for the C2 layout gave overpredictions instead with a much-increased V_G. These results indicate considerable dependency on the layout of solid wall boundaries.

The results from the two studies spanning a decade are consistent with each other in that C1 was consistently underpredicted with a variance in the mid to low single digits, and C2 was consistently overpredicted and with a very large variance.

An MV diagram showing a summary of these results is given in Figure 6.

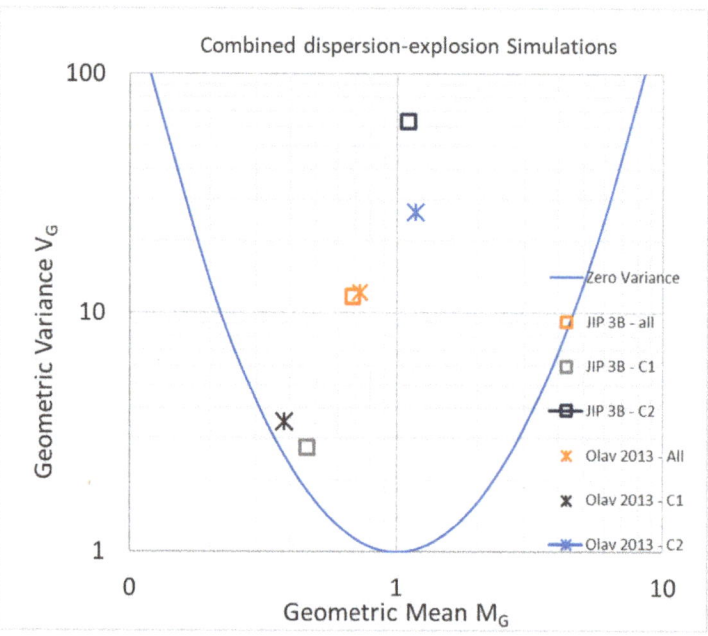

Figure 6. An MV diagram summarising the performance of the 'direct dispersion-explosion simulations' using FLACS. No ESC was assumed. Non-homogeneous flammable gas clouds were calculated and used directly as input to explosion simulations. Results from [3] in crosses and the Phase 3B reports [17] in hollow squares. Note about legend: a line is added to the start of a new symbol to allow its easy identification.

Effect of Control Volume Sizes—Grid Dependency for Dispersion-Explosion Linked Simulations

The effect of grid sizes on the simulation results is significant. Data issued by GexCon as part of the Phase3b JIP showed results for three grid sizes: 0.5 m, 1 m and 1.33 m; these are grid sizes outside the grid refinement zone. The refinement zone is the region immediately around the gas jet in order to obtain more accurate estimation of jet behavior through resolving flow details better. This zone contains a range of grid sizes smaller than those outside this region. A summary of the results is given in Table 2.

Table 2. Geometric Mean and Variance: direct dispersion–explosion simulations as a function of grid sizes. Results taken from the Phase 3B joint industry project (JIP) [17]. The results are based on maximum overpressure within the module (in experiments and in simulations).

Grid Sizes	0.5 m		1 m		1.33 m	
	M_G	V_G	M_G	V_G	M_G	V_G
All	0.68	11.7	0.54	18	0.30	32
C1 only	0.46	2.73	0.34	15	0.21	25
C2 only	1.10	63.6	1.0	22	0.45	42

The results show significant dependency on grid sizes, see Figure 7. As grid size increases, the bias worsens; M_G values move from 0.68 to 0.54 to 0.3 for grid sizes of 0.5 m, 1 m and 1.33 m, respectively. Variance also increases with grid sizes indicating that the model results become increasingly unreliable. The two subsets of data for the C1 and C2 layouts show similar trend with higher values of variance.

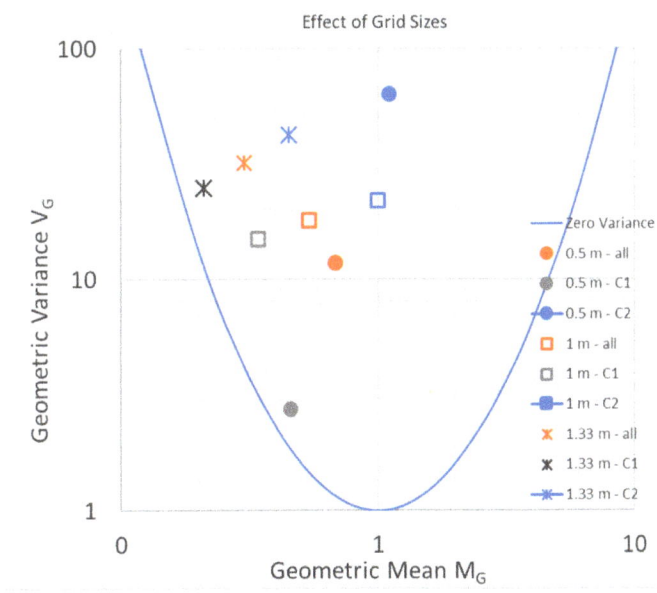

Figure 7. Results of direct dispersion–explosion simulation using three grid sizes from the Blast and Fire for Topside Structure joint industry (BFTSS) Phase 3B project report. Grid size refers to grids outside the local refinement region round the release source. Note about legend: a line is added to the start of a new symbol to allow its easy identification.

The results shown in Table 1 indicated that the results of Hansen 2013 were mostly likely based on grids of order 0.5 m.

6.2. The ESC Model—Q9

The methodology in explosion hazard assessment using the Q9 model involves two steps: (a) calculating a flammable volume and (b) locating the Q9 gas cloud in various locations for the purpose of determining maximum overpressures.

6.2.1. Cloud Volumes Derived from Experimental Data

Results are summarised in Table 3 which gives the maximum calculated overpressure of any of the sensor locations against maximum measured overpressures of any sensors. This was for Q9 volumes derived from experimental gas concentration measurements.

Table 3. Cloud located at locations conducive to generating high overpressures: Geometric Mean and Variance. Maximum P column taken from [2].

	Maximum P		Average P	
	M_G	V_G	M_G	V_G
All	1.0	4.6	0.9	3.7
C1 only	0.7	1.6	0.7	1.7
C2 only	1.7	23	1.6	12

Table 3 also contains additional results on averaged overpressures. They give the averaged values of all maximum pressures of all sensors compared with the average of maximum pressures calculated at all sensor locations within the module. Generally averaged values have marginally smaller M_G and smaller variance; this is as expected.

As stated in the Tam 2008 paper, the results showed that when all the data are considered, the Q9 model is neutrally biased. This is reflected in the results in Table 3.

The results in Table 3 are summarised in Figure 8. Included in the figure is also the results of placing the Q9 clouds near the edge of the module. This will be discussed later.

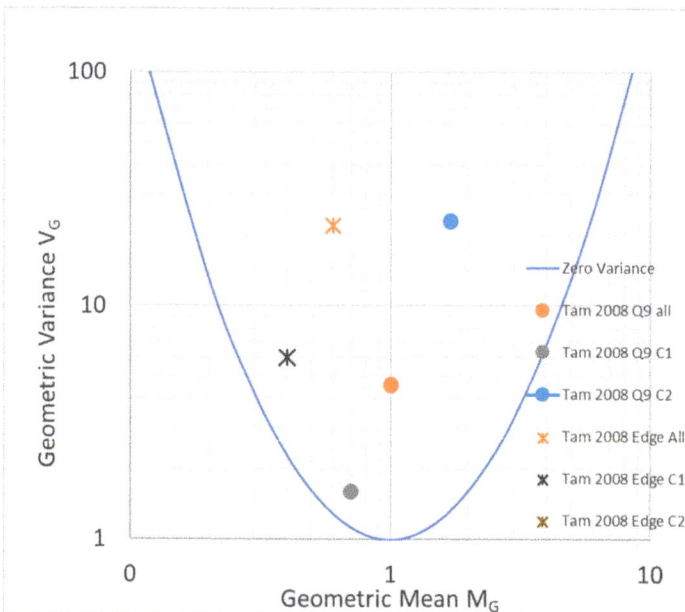

Figure 8. MV diagram showing performance of the Q9 model based on work carried out towards the Tam 2008 paper. The solid circles show results in which Q9 volumes were derived directly from experimental data (published in [2]). The crosses are results in which Q9 volumes are placed near the edge of the module. The variance of the blue cross is so large that it is outside the boundary of this figure. Note about legend: a line is added to the start of a new symbol to allow its easy identification.

6.2.2. Cloud Volumes Derived from FLACS Dispersion Simulations

This reflects how ESC is used in ERA in practice. Q9 volumes are calculated by FLACS which sums up Q9 volume contributions from every grid cell within pre-defined boundaries of the area of interest (e.g., a compression module on a platform).

This work was carried out in 2019. As previous work indicated sensitivities to grid sizes, we decided to calculate overpressures using two grid sizes: 0.5 m and 0.25 m. It would have been necessary to apply more than these two grid sizes in practice during an ERA due to the range of ESC volumes being considered. A summary of the results is given in Table 4.

Table 4. Geometric means and variance: Q9 Cloud volumes calculated by FLACS was used as an input to explosion simulations.

	Grid Size—0.5 m		Grid Size—0.25 m	
	M_G	V_G	M_G	V_G
All	0.6	13.7	0.6	7.5
C1 only	0.3	14.3	0.4	7.7
C2 only	1.5	12.7	1.2	7.2

The trend of results of "All", "C1" and "C2" are consistent in all the results above, in that "C1 only" results are underpredicted and "C2 only" results are overpredicted. Over the entire dataset, the Q9 model led to underprediction. Specifically, on this set of results, it can be seen that the smaller grid size of 0.25 m gave better quality results: the range of bias is reduced, and the magnitude of variance is nearly halved.

Figure 9 summarises the results.

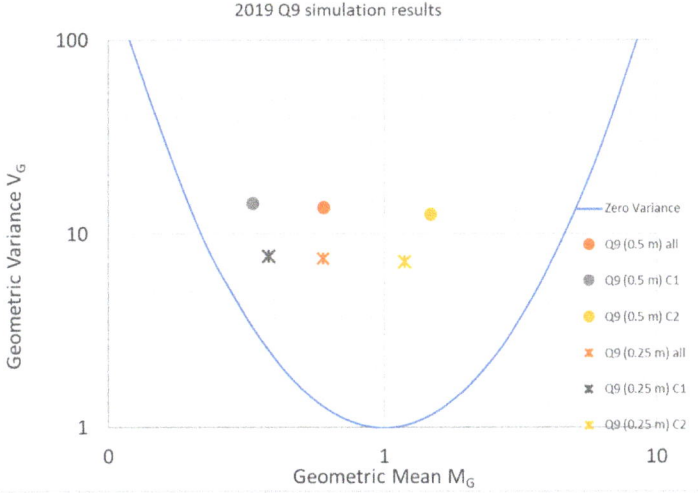

Figure 9. MV diagram showing the results using two grid sizes (0.5 m and 0.25 m) using current explosion risk assessment (ERA) methodology.

6.3. Sensitivity to Cloud Locations

Predicted overpressures are sensitive to where the Q9 cloud is placed. This was discussed in [2] in which authors compared calculated overpressures of ESCs placed at the centre of the test module with those located at the edge. The average ratio was found to be about three.

The sensitivity to location is shown by the associated variance V_G which has a higher value than those sited in locations for highest overpressures (see Figure 8).

Results indicates that this sensitivity tends to be higher for the ESC model that produces smaller cloud volumes than those which produce larger volumes. For example, the ESC model, ">FL" or ΔFL, produces larger cloud volumes than those from the Q9 model; they show less sensitivity [2].

7. Discussions

7.1. Consistency of Results

All the results calculated over the last two decades using various methods of deriving Q9 volumes have the same behaviour: Q9 underpredicts in all of them when the whole Phase 3B dataset is used, and the trend in predictions for the two layout configurations is consistent.

7.2. Sensitivity of FLACS to Boundary Properties

While this is not the main objective of this paper, the Q9 results show that there is a strong sensitivity to layout of boundary walls. In the two configurations in Phase 3B, the C1 configuration (see Figure 3) is close to a "tunnel" type layout with two parallel walls and the prevailing wind direction being along the length of the module, while the wall arrangement in C2 is a U shape with the open U on the long side. Wind direction was always along the length of the experimental module. In the C1 layout, wind flowed through the two open ends of module (Figure 3 bottom right). In the C2 layout, wind blew across the open side (Figure 3 bottom left). The results strongly suggest that as the wall arrangement deviates from a tunnel arrangement, the Q9 model becomes more unreliable as shown by the large value of variance V_G.

7.3. Reliability of Direct Dispersion–Explosion Simulations

This method suffers from excessive variance, and sensitivities to grid sizes and wall layout.

Like all CFD codes, FLACS is susceptible to grid size sensitivity. GexCon has made significant effort to limit its effect, e.g., there is a firm guidance on grid sizes for explosion simulations. However, the guidance for direct dispersion–explosion simulations does not exist. As results here show, this direct dispersion–explosion methodology performs poorly and is prone to grid sensitivities. As it is, there is no justification of using this method: it is less accurate and more costly.

Further validation and guidance are clearly needed.

7.4. Q9 Is Not Conservative

Hansen et al [3] asserted repeatedly that the Q9 model is conservative and that all other methods are excessively conservative. They imply that Q9 is conservative through an indirect comparison with data. However, their own conclusion did not support their own assertion: "Pressures achieved when exploding quiescent Q9 based ideal clouds correlate quite well to experimental pressures from ignited jets, but for some test scenarios higher pressures are seen than obtained with the quiescent ideal clouds". At best, Their own conclusion indicates that the Q9 model is neutrally biased.

Results of this paper and [2] do not support the "over conservative" assertion.

7.5. Complexity and Accuracy

We found that many of these assertions on "conservatism" are based on theoretical argument, sensitivity calculations or "experience". The misconception is that a method or a mathematical model containing some physics must be more accurate and better than methods containing less. The devil is in the physics that are missing. Complexity should not be confused with accuracy. This was discussed in [2].

7.6. Key Learning

7.6.1. Recommended Usage of Q9 Is Inadequate

Hansen et al.'s recommendation is that only the calculated maximum overpressure values (over the entire volume of interest and across all ESC locations) are used for probabilistic assessment for each release scenario (e.g., producing exceedance curve): this approach would be conservative. It could be. Only if Q9 is unbiased.

However, results show that the Q9 model systematically underpredicts, hence the assertion of conservatism does not hold. Secondly, this recommendation is not universally applied in practice in ERAs. We will discuss the effect of conservatism or lack of it next.

7.6.2. Applications of ESC in Probabilistic Explosion Analysis to Exceedance Curve

It is common to use a risk-based approach to define explosion loads for structural design. The design accidental load (DAL) is an example of this. DAL is widely used, and its value is derived from an exceedance curve. DAL is defined as the explosion load on a structural element for a specific cumulative risk frequency which can be obtained from an exceedance curve. This is based on the limit state approach in the design of structures against gas explosions [28,29]. This approach was based on an established methodology on design against extreme environmental loads such as earthquakes and ocean waves. This is now encapsulated in the Fire and Explosion Guidance [18]. It recommends that design loads correspond to two limit states, strength level blast (SLB) for frequent, and ductility level blast (DLB) for rare events, respectively: the former ensures integrity which minimises structural and equipment damage that could lead to prolonged operation disruption, the latter for protection of people through avoidance of progressive structural collapse. This guidance is consistent with that in NORSOK Z-013 and ISO 19901-3. The threshold of cumulative frequency for DLB varies. The guidance defined a minimum of 10^{-4}/yr, and some companies chooses the other lower bound, e.g., 10^{-5}/yr in order to keep total risks from both fire and explosion to below 10^{-4}/yr.

The effect of systematic biases of an ESC model can be significant. For example, the DAL derived from exceedance curves or explosion risk analysis could be underestimated if the ESC is systematically underpredicting.

This can be illustrated by comparing exceedance curves derived from two ESC models. This example is taken from a study that we undertook on an existing facility. We calculated the DAL values for a structural wall using the two ESC models, ΔFL and Q9. Figure 10 shows the two exceedance curves derived from them for a wall on the living quarter. It can be seen that there are significant differences between them. The key points illustrated in the figure are:

(i) The DAL derived from the Q9 model is more than three times smaller than that from the ΔFL model, e.g., 0.4 bar and 1.3 bar, respectively at a cumulative frequency of 10^{-5}/yr. The magnitude of DAL varies according to the level of exceedance set. At 10^{-4}/yr. The magnitude of overpressure is much lower, though the ratio remains similar.

(ii) The maximum overpressure calculated also varies about three-fold: 1.6 bar compared with 0.6 bar for the ΔFL and Q9 models, respectively. This higher maximum overpressure is probably not used for structural design of the blast wall in a risk-based approach. However, there are impacts on hazard management, e.g., on (a) emergency planning and management and (b) siting and layout particularly for onshore facilities at an early stage of design, e.g., the concept selection stage.

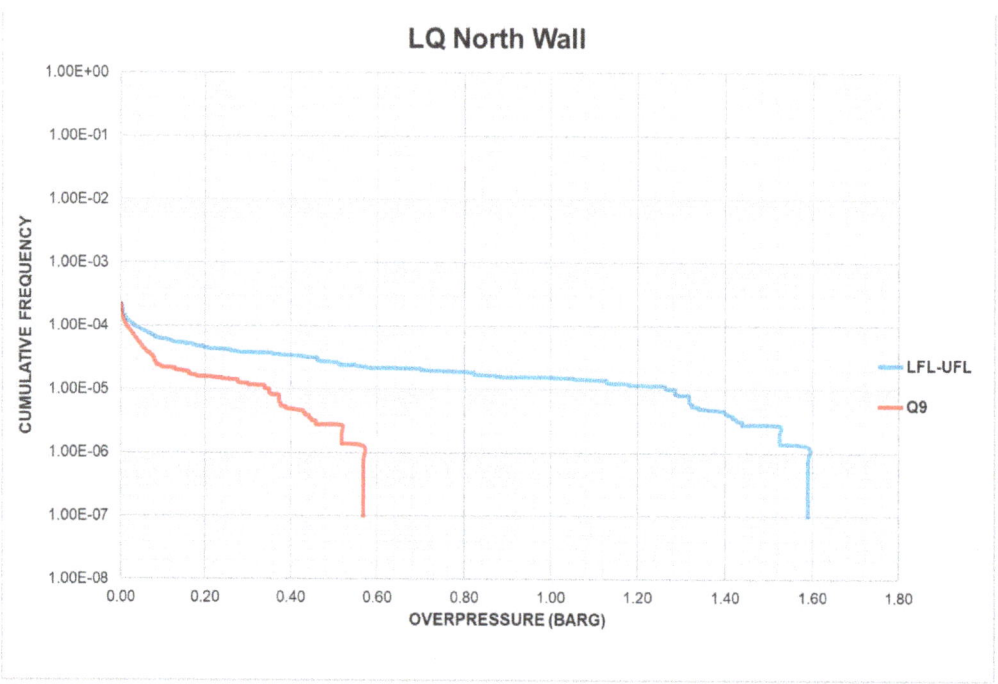

Figure 10. Exceedance curve showing cumulative frequencies for overpressures on a structural wall of a living quarter taken from A design study of an offshore production facility. These curves were derived from ESC model: Q9 and ΔFL (concentration between upper and lower flammability limits, this is labelled as "LFL-UFL" in the legend).

The results presented serve to highlight the choice of an ESC model can have a significant impact on design and risk management. An ESC model that systematically underpredicts will also underpredict DAL.

This could lead to inadequate design of blast walls and other critical structures.

7.6.3. Recognised Evaluation and Presentation Format

When presenting results, a recognised and common format should be used. Over-complex figures with lots of data points are confusing, not to mention there are many data points which overlap, and true results are obscured.

A common recognised format should be used. The MV diagram is used in MEGGE [19] and in the model evaluation exercises in the BFTSS Phase 2 project. MV diagram is deployed in all model evaluation protocols [16,21].

It is worth noting that, when results of the two papers ([2] and [3]) are analysed and presented on MV diagrams, there is no inconsistency between their results. This highlights the importance of using recognised evaluation protocol to avoid confusion.

7.6.4. An Explosion Risk Assessment in Practice

The current guidance for explosion modelling in FLACS requires that there be a minimum number of control volumes across the minimum dimension of an ESC. As some of the Q9 volume can be small, two issues arise:

(a) Adherence to guidance: this would require carrying out simulations using different grid sizes depending on results of dispersion calculations. As a typical ERA involves large numbers (of order a 1000) of simulations, there is a temptation to bypass or approximate this step given time and cost pressure.

(b) The effect of mixing or using different grid sizes is difficult to quantify—as can be seen above, FLACS is grid sensitive. The net effect of mixing results obtained using various grid sizes has not yet been quantified.

7.6.5. Quality Control Issues

Best and common practice implementation across the consultant industry is an issue. Our experience from reviewing reports from a range of consultants is that rigorous quality assurance by company or independent experts is necessary as ERA processes and procedures involved in the application of the Q9 model vary and the impact of this difference is not obvious and usually not stated. This issue is not new [30]. A recent JIP proposal by the UK Health and Safety Laboratory goes some way to address it [15].

7.7. More Recent Data on Hydrogen

The Q9 and other ESC models described here were developed for hydrocarbons. The burgeoning of hydrogen economies would demand similar assessments as those for the hydrocarbon economy. Application of any ESC model for hydrogen should be verified by data from full scale hydrogen experiments.

Data from Phase 3B are not appropriate.

Current "validation" results based on hydrocarbon cannot be carried forward to applications to hydrogen. Appropriate hydrogen data should be used to calibrate ESC models. We understand that there are recent and ongoing research programmes which provide near full scale data for specific applications (e.g., on hydrogen filling stations, [31,32]).

7.8. Application of Q9 in Reality

The comparison so far is based on maximum values among all sensors in the experiment with maximum calculated values selected from those sensor locations, i.e., the maximum of all sensors against the maximum of all monitoring points. A monitoring point is a location in the calculation domain where values of fluid properties are extracted or "monitored". In this case, the monitoring point locations duplicate those in the experiment. The recommendation of the use of the Q9 model is aligned with the validation methodology. That is, only the maximum calculated values from all possible locations of Q9 clouds are used.

This is akin to using CFD models as though they are simple empirical or phenomenological box models which are much simpler and cheaper to run.

The issue is a philosophical one. What is the point of using a complex and expensive code when you cannot use all the other results generated? The temptation to utilise volumes of detailed calculation results is overwhelming, particularly for the uninitiated. Unsurprisingly, more and more complicated analyses are developed, utilising the vast amount of calculated values for other parameters. We will explore some of these analyses later.

7.9. Expectation and Reality—Application Outstripping Capability

Expectation of the capability of commercial explosion codes like FLACS is high. There is a danger that expectation outstrips the capability that can be realistically delivered. We will explore some examples with the application of the Q9 model.

In common with other CFD codes, FLACS produces many details, showing distribution of pressure in space and in time as well as other useful parameters such as velocity, drag forces, etc. The availability of this information often leads to changes in assessment methodologies impacting on design. While published validations have been on global values (e.g., maximum (or mean) among measurements) of a limited set of model outputs, there is no published systematic validation exercise on their spatial and temporal distribution that we are aware of.

Our observation is this. The mere fact that output of these other model parameters is available will lead to their use in practice. When one aspect of the code is deemed to be validated, it is then assumed that the whole code is validated covering other aspects beyond the limited validation exercise.

Here, are two examples to illustrate the point.

7.9.1. Refinement of Transient Nature of Releasees

As large releases tend to be short-lived due to finite inventories, releases from large leaks tend to reduce with time, e.g., as a blow down system is activated or inventory depleted. There is an increasing use of the time-varying characteristics of the transient nature of hydrocarbon releases to define a more "accurate" risk value. This is done by integrating instantaneous risk values over the period of hydrocarbon release. The incorporation of the transient effect usually results in lower risk figures than steady release over the release period.

The integrated risk values are sensitive to the choice of ESC model. Q9 cloud volume–time history has a common characteristic in that with an initial peak the cloud volume decays more or less monotonically with time (see Figure 11). Other ESC models may not share this. This is illustrated by comparing ESC volumes from two models: Q5 (Q9 and Q5 are very similar in definition (see Figure 2) and Q0 from the FLACS simulation of Test 13 in Phase 3B [17], see Figure 12.

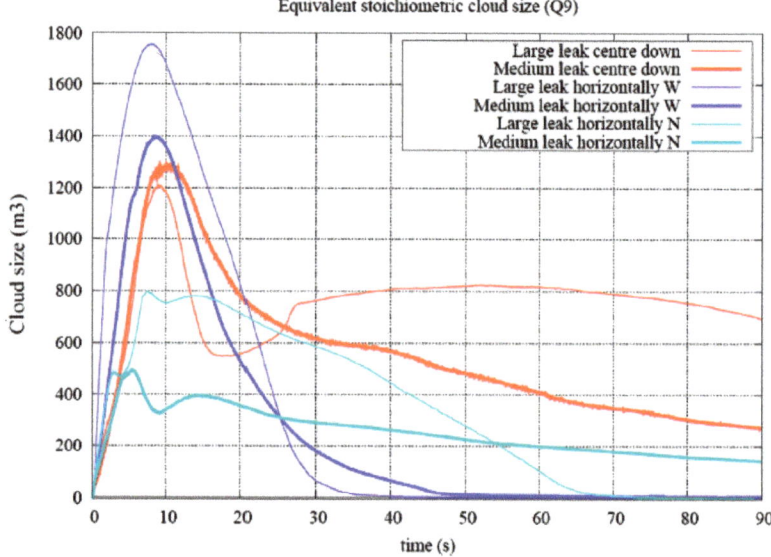

Figure 11. Cloud size vs. time profiles for Q9 clouds (taken from [3]). They show characteristic flammable cloud volume variation with time.

The evolution of these two ESC volumes with time is very different. In this example, the Q9 model would predict an integrated risk of more than half that derived from Q0 assuming uniform ignition probability with time.

We highlighted this aspect here because the maximum volume and subsequent evolution of flammable volume depend on the choice of the ESC model. There has not been a study on the validity of any ESC model for this application.

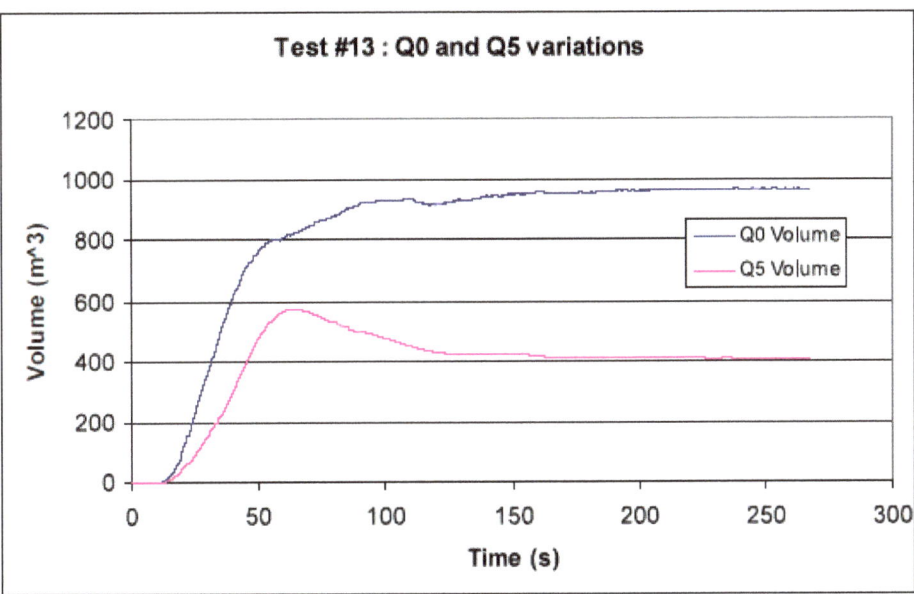

Figure 12. Calculated evolution of ESC volumes from experimental data for two ESC models: Q0 and Q5. This shows the sensitivity of the ESC model on maximum flammable cloud volumes and their evolution with time (taken from [17]).

7.9.2. Advanced CFD/FE Analysis

There is an increasing trend to make use of detailed output from FLACS to "optimise" structural design. The spatial and temporal distribution of pressure results are extracted, and this is applied as direct input to a finite element structural response code so that the temporal and spatial characteristics of the load are accounted for in the structural response.

If there is validation of model predictions on spatial distribution with time and the performance is good, the adoption of this approach would have been compelling. However, this validation has not been carried out. There were some limited study for theoretical worst-case scenarios on coupled loading–response analysis [11,33].

Our analysis shows that there is a large variance (V_G) associated with model predictions, as shown above in Section 6.1. Figure 13 is a comparison of predicted and observed maximum overpressure at all sensor locations for Test 17. We chose this test because the Q9 model gives the best match between data and simulation results. The figure shows that there is a large variation between predicted and observed maximum overpressures at each of these sensor locations.

Our results here show that the variance of predictions against measurements for individual points is much higher than those of global maximum against global maximum or global average against global average results. It is possible that these variances when combined (with uncertainties associated with structural response analysis) all cancel out to provide an accurate or acceptable description of response. This has yet to be demonstrated.

We contend that the approach of mapping spatial and time distribution results onto structural FE code is susceptible to unquantifiable error. We suggest that development work and guidance on this application are needed.

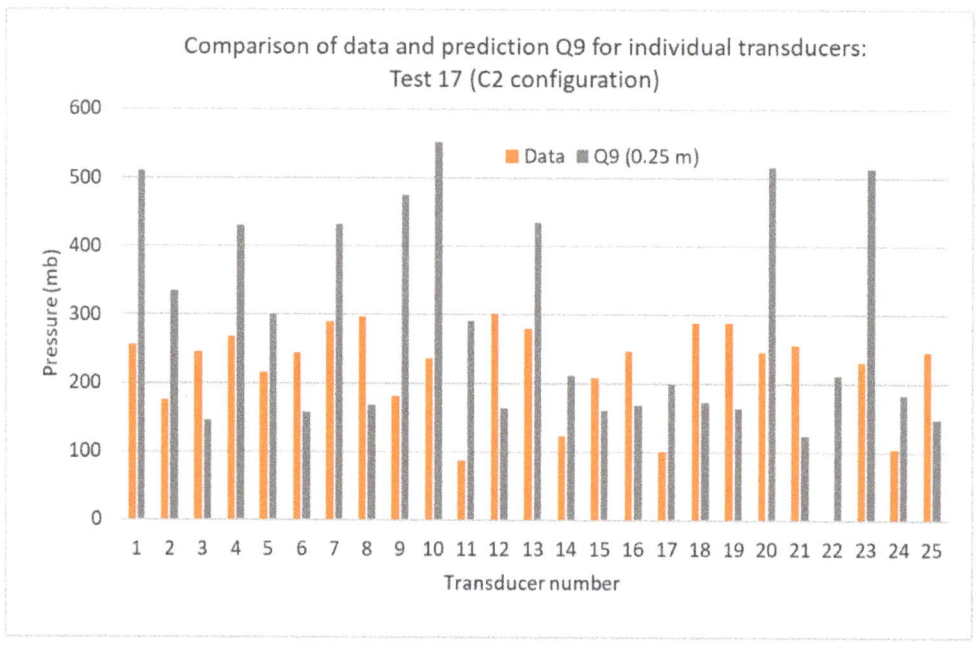

Figure 13. Comparison of predicted and observed maximum overpressures for all transducers in Test 17 in the Phase 3B JIP. The predicted values were calculated from FLACS using Q9 ESC and a grid size of 0.25 m.

8. Conclusions

A comprehensive review was carried out on previous work (from 2001) on comparison between data from the large-scale realistic release JIP (Phase 3B) and simulation results of FLACS using the Q9 model.

The Q9 model in FLACS significantly underpredicts the Phase 3B dataset. The value of the geometric mean (M_G) is 0.6 for maximum overpressures. This has the effect of underestimating design accidental load, leading to inadequate design or underestimation of explosion risks.

Results also show that there is a large systematic difference in the Q9 model performance for the two wall layouts in Phase 3B—underprediction and smaller variance in the tunnel geometry compared with overprediction and higher variance in the U shape geometry.

9. Observations

- There is an assumption that the Q9 model is more accurate because it contains more physics than one containing less. This is not true. The devil is in the missing physics. The true arbiter is validation with full scale data.
- Recognition is lacking in the industry of the limitation of the code, and the lack of appropriate data and validation. As a result, ever more complex analyses are being carried out where there is no validation or detailed scientific evaluation to support them (e.g., time varying risk calculation, time and spatially coupled-load–response analysis on structures, etc.).
- Recognised model evaluation protocols should be used in all publications on model validation or model evaluation. This provides clarity and avoid confusion.
- The hydrogen economy is fast developing. The application of the Q9 model to hydrogen requires a separate evaluation/validation exercise based on hydrogen data derived from experiments at appropriate scale.

10. Recommendations

This study highlighted five main areas where further work is recommended:
(a) Develop an appropriate ESC model through vigorous scientific methods and validated by appropriate scale data.
(b) Q9 should be used with caution; a full account of systematic underprediction and large variance should be made.
(c) Generate appropriate large-scale data with a wider range of conditions than those of Phase 3B, e.g., involving a range of geometries and boundary wall arrangements, and non-steady scenarios, such as simulation of blow downs and delayed ignition. The industry should form a JIP to address this.
(d) The issue of variability due to users in the execution of ERA, particularly involving the Q9 model should be addressed. A JIP that goes someway to address this issue was suggested by the Health and Safety Laboratory on a joint ERA inter-comparison exercise involving key consultants [15].
(e) The validity and applicability of coupled explosion loading/structural response analysis should be clarified. This could involve a guidance to define correct application, evaluation against data to establish confidence limits, etc.

Author Contributions: Conceptualization, V.H.Y.T.; methodology, F.T., V.H.Y.T. and C.S.; formal analysis, F.T., C.S.; investigation, F.T., V.H.Y.T., C.S.; resources, C.S., F.T.; data curation, F.T.; writing—original draft preparation, V.H.Y.T.; writing—review and editing, F.T., C.S.; visualization, F.T., C.S.; supervision, V.H.Y.T.; project administration, V.H.Y.T. All authors have read and agreed to the published version of the manuscript.

Funding: This research received no external funding.

Institutional Review Board Statement: Not applicable.

Informed Consent Statement: Not applicable.

Data Availability Statement: Not applicable.

Conflicts of Interest: The authors declare no conflict of interest.

Abbreviations

ΔFL	Concentration between upper and lower flammability limits.
> LFL	Concentration above the lower flammability limit.
M_G	Geometric mean.
V_G	Geometric variance.
BFTSS	Blast and Fire for Topside Structure joint industry—the name of a series of JIPs spanning over more than a decade. There are four phases: 1, 2, 3a and 3b.
CFD	Computation fluid dynamics.
CMI	Christian Michelsen Institute.
DAL	Design accidental load.
ERA	Explosion risk assessment.
ESC	Equivalent stoichiometric cloud. This refers to dispersion-based ESC where not specifically stated in the text.
FE	Finite Element
JIP	Joint industry project.
LFL	Lower flammability limit.
MEGGE	Model evaluation group for gas explosion.
MV diagram	Mean–variance diagram
Phase 3B	One of the phases of the BFTSS project which addressed realistic release scenarios
Q9	One of the widely used models of ESC. Its definition is in 2.
UFL	Upper flammability limit.

References

1. Norwegian Oil Industry Association. *Risk and Emergency Preparedness Assessment—Norsok Standard Z-013*; Norwegian Technology Centre: Oslo, Norway, 2001.
2. Tam, V.H.Y.; Wang, M.; Savvides, C.N.; Tunc, E.; Ferraris, S.; Wen, J.X. Simplified flammable gas volume methods for gas explosion modelling from pressurized gas releases: A comparison with large scale experimental data. In *Institution of Chemical Engineers Symposium Series*; IChemE: London, UK, 2008.
3. Hansen, O.R.; Gavelli, F.; Davis, S.G.; Middha, P. Equivalent cloud methods used for explosion risk and consequence studies. *J. Loss Prev. Process. Ind.* **2013**, *26*, 511–527. [CrossRef]
4. HMSO. *The Flixborough Disaster: Report of the Court of Inquiry*; HMSO Stationery Office: London, UK, 1975.
5. Prörtner, H. Gas cloud explosions and resulting blast effects. *Nucl. Eng. Des.* **1977**, *41*, 59–67. [CrossRef]
6. Gugan, K. *Unconfined Vapour Cloud Explosions*; Institution of Chemical Engineers: London, UK, 1979.
7. Berg, A.V.D. The multi-energy method: A framework for vapour cloud explosion blast prediction. *J. Hazard. Mater.* **1985**, *12*, 1–10. [CrossRef]
8. Mercx, W.P.M.; van den Berg, A.C.; van Leeuwen, D. *Application of Correlations to Quantify the Source Strength of Vapour Cloud Explosions in Realistic Situations. Final Report for the Project: 'GAMES'; TNO Report PML 1998-C53*; Netherlands Organization for Applied Scientific Research (TNO): The Hague, The Netherlands, 1998; p. 156.
9. Cullen, W.D. *The Public Inquiry into the Piper Alpha Diasater*; HMSO: London, UK, 1990; Volume 1–2.
10. Tam, V.H.Y.; Langford, D. Design of the BP Andrew platform. In *International Conference on Health, Safety and Environment in Oil and Gas Exploration and Production*; ERA Technology: London, UK, 1994; pp. 3.6.1–3.6.10.
11. Selby, C.; Burgan, B. *Blast and Fire Engineering for Topside Structures Phase 2*; Final Summary Report; The Steel Construction institute: London, UK, 1998.
12. Al-Hassan, T.; Johnson, D.M. Gas explosions in Large Scale Offshore Module Geometries: Overpressures, Mitigation and Repeatability. In Proceedings of the International Conference on Offshore Mechanics and Arctic Engineering—OMAE 98, Lisbon, Portugal, 5–6 July 1998.
13. Shearer, M.J.; Tam, V.H.; Corr, B. Analysis of results from large scale hydrocarbon gas explosions. *J. Loss Prev. Process. Ind.* **2000**, *13*, 167–173. [CrossRef]
14. Cleaver, R.P.; Britter, R.E. A workbook approach to estimating the flammable volume produced by a gas release. *FABIG Newsl.* **2001**, *30*, 5–7.
15. Stewart, J.R.; Gant, S. A Review of the Q9 equivalent cloud method for explosion modelling. *FABIG Newsl. March* **2009**, *75*, 16–25.
16. Coldrick, S. How do we demonstrate that a consequence model is fit-for-purpose? *Inst. Chem. Eng. Symp. Series* **2017**, *162*, 1–9.
17. Johnson, D.M.; Cleaver, R.P. *Gas Explosions in Offshore Modules Following Realistic Releases (Phase 3B): Final Summary Report*; Advantica: British Gas, Loughborough, 2002.
18. Oil and Gas UK. *Fire and Explosion Guidelines—Issue 2*; Oil and Gas UK: London, UK, 2018.
19. Selby, C.; Burgan, B. *Explosion Model Evaluation Project*; Report RT 738; The Steel Construction Institute: London, UK, 1998.
20. Hanna, S.R.; Strimaitis, D.G.; Chang, J.C. Evaluation of fourteen hazardous gas models with ammonia and hydrogen fluoride field data. *J. Hazard. Mater.* **1991**, *26*, 127–158. [CrossRef]
21. Ivings, M.; Lea, C.; Webber, D.; Jagger, S.; Coldrick, S. A protocol for the evaluation of LNG vapour dispersion models. *J. Loss Prev. Process. Ind.* **2013**, *26*, 153–163. [CrossRef]
22. Tam, V.H.Y. Explosion model evaluation. *FABIG Newsl.* **1998**, *22*, 34–36.
23. Bowerman, H.; Owens, G.W.; Rumley, J.H.; Tolloczko, J.J.A. *Interim Guidance Notes for Design and Protection of Top-Side Structures against Explosion and Fire*; SCI-P-112; The Steel Construction Institute: London, UK, 1992.
24. Cleaver, R.P.; Burgess, S.; Buss, G.Y.; Savvides, C.; Tam, V.H.Y.; Connolly, S.; Britter, R.E. Analysis of Gas Build-Up from High Pressure Natural Gas Releases in Naturally Ventilated Offshore Modules. In *8th Annual Conference on Offshore Installations: Fire and Explosion Engineering*; ERA Technology: London, UK, 1999; pp. 1–12.
25. Savvides, C.; Tam, V.H.Y.; Kinnear, D. Dispersion of Fuel in Offshore Modules: Comparison of Predictions Using Fluent and Full Scale Experiments. In *Major Hazards Offshore Conference*; ERA Technology: London, UK, 2001; pp. 1–14.
26. Savvides, C.; Tam, V.H.Y.; Os, J.E.; Hansen, O.; Wingerden, K.; Van Renoult, J. Dispersiion of Fuel in Offshore Modules: Comparison of Predictions Using FLACS and Full Scale Experiments. In *Major Hazards Offshore Conference*; ERA Technology Report: London, UK, 2001; pp. 1–8.
27. Chang, J.C.; Hanna, S.R. Air quality model performance evaluatione. *Meteorol. Atmos. Phys.* **2004**, *87*, 167–196.
28. Tam, V.H.; Corr, B. Development of a limit state approach for design against gas explosions. *J. Loss Prev. Process. Ind.* **2000**, *13*, 443–447. [CrossRef]
29. Corr, R.B.; Frieze, P.A.; Tam, V.H.Y.; Snell, R.O. Development of the Limit State Approach for Design of Offshore Platforms. In Proceedings of the Conference on Fire and Explosion Engineering, ERA Technology, Leatherhead, UK, December 1999.
30. Holen, J. Comparison of five corresponding explosion risk studies performed by five different consultants. In *Major Hazards Offshore Conference Proceedings*; ERA Technology: London, UK, 2001.
31. Wen, J.; Rao, V.M.; Tam, V. Numerical study of hydrogen explosions in a refuelling environment and in a model storage room. *Int. J. Hydrog. Energy* **2010**, *35*, 385–394. [CrossRef]

32. Tanaka, T.; Azuma, T.; Evans, J.; Cronin, P.; Johnson, D.; Cleaver, R. Experimental study on hydrogen explosions in a full-scale hydrogen filling station model. *Int. J. Hydrog. Energy* **2007**, *32*, 2162–2170. [CrossRef]
33. Caretta, G. Coupled Modelling and Experiments in Structural Response to Blast Loading. Ph.D. Thesis, University of Cambridge, Cambridge, UK, 2004.

Article

Influence of the Precoat Layer on the Filtration Properties and Regeneration Quality of Backwashing Filters

Volker Bächle *, Patrick Morsch *, Marco Gleiß and Hermann Nirschl

Karlsruhe Institute of Technology (KIT), Strasse am Forum 8, 76131 Karlsruhe, Germany; marco.gleiss@kit.edu (M.G.); hermann.nirschl@kit.edu (H.N.)
* Correspondence: volker.baechle@kit.edu (V.B.); patrick.morsch@kit.edu (P.M.); Tel.: +49-721-608-42427 (V.B. & P.M.)

Abstract: For solid–liquid separation, filter meshes are still used across large areas today, as they offer a cost-effective alternative, for example, compared to membranes. However, particle interaction leads to a continuous blocking of the pores, which lowers the flow rate of the mesh and reduces its lifetime. This can be remedied by filter aids. In precoat filtration, these provide an already fully formed filter cake on the fabric, which acts as a surface and depth filter. This prevents interaction of the particles to be separated with the mesh and thus increases the service life of the mesh. In this work, the influence of a precoat layer with different fibre lengths of cellulose on the filtration behavior is investigated. A satin with a pore size of 11 μm is used as the filter medium. The effects of the precoat layer on the filter media resistance, the filter cake resistance, the turbidity impact, and the regenerability of the fabrics are investigated. This study shows an overview of the suitability of various cellulose fibres based on different aspects as filter aids for particles in ultrafine filtration.

Keywords: precoat layer; precoat filtration; cellulose fibres; filter aids; backwash filtration; filter regeneration; filter media resistance; filter cake resistance; turbidity; particle layer

1. Introduction

For filtration of suspensions with fine particulate solids through filter cloth, it is necessary, due to the high pressure drop of such aggregates, to carry out filtration at low filter cake thicknesses or high filtration pressures. In these cases, an economical filtrate volume flow is ensured. In the macroscopic description of the filtration process according to the filter equation (Equation (1)), the filtration pressure Δp is described as a function of the sum of the filter cake resistance $\alpha_H \cdot H_{FC}$ and filter mean resistance R_{FM}. Furthermore, the dynamic viscosity η_f, the filtrate volume flow $\dot{V}_F$, and the filter area A represent variables that influence the filtration pressure Δp. As in the macroscopic, the microscopic process of particle separation in and on the meshes of the filter fabric is decisive for the quality of the filtration. If the mesh of the fabric is larger than the diameter of the particle system to be filtered, particle penetration (turbidity blow) occurs, and the quality of the filtrate suffers as a result [1]. The more unbalanced the ratio in the direction of "large mesh, fine particles", the more pronounced the turbidity impact. In the worst case, no filtration takes place at all and no filter cake is formed on the mesh.

$$\Delta p = (\alpha_H \cdot H_{FC} + R_{FM}) \cdot \eta_f \cdot \frac{\dot{V}_F}{A} \qquad (1)$$

One possible option for improving filtration is to replace the filter fabric with a finer woven mesh. This is also recommended in view of the large amounts of particle penetration in the filtrate. Scientific studies already provide good reference values as to which meshes are suitable for the filtration of a wide variety of particle systems [2–5]. Furthermore, the choice of mesh is often based on practical experience and the know-how of the equipment operators, manufacturers, and mesh suppliers. According to [6], the mesh selection is characterized independently of manufacturers as a function of mesh size, particle size distribution, and particle shape. These are based on experiments with the pressurized filter cell, a reliable instrument for selecting suitable meshes [7]. Satin fabrics in particular have a smooth surface due to their weave and are advantageous for applications with backwashing and clogging behaviour. However, they have a comparatively higher filter medium resistance, which is why it is a trade-off between high filter resistance and advantageous backwashing. Therefore, tests are necessary for the individual case [8]. In addition to selecting a more suitable mesh, the addition of filter aids is also an option. These aids form a "particle protection layer" directly on the filter mesh, thus allowing more extensive filtration of the particle system [9,10]. The particles are then deposited on the surface of the filter aid and in its porous structure.

In this paper, the influence of filter aids during filtration is investigated in more detail. For this purpose, three filter aids are investigated and their effect on filtration with three selected particle systems is described according to [8,11]. This is done on a selected number of fabrics according to [6]. In addition to the effect of a precoat layer on the filtration, an analysis of the regeneration behavior in the backwashing filtration is also carried out. For this purpose, a flow reversal of the filtrate stream occurs, which detaches the filter cake together with the precoat layer from the tissue and throws it off. This is done in a liquid environment according to [8]. The results of these investigations can be applied in the filtration of fine particle systems with filter aids, including subsequent regeneration.

2. Theory

The use of filter aids for filtering liquids is a widespread field of application and can be found, for example, in the clarification of beer and wine (food industry), but also in the chemical industry [12]. The effect of the precoat layer and the objective are always identical: the addition of a further particle system with a different particle size distribution facilitates filtration because the looser structure of this added particle structure leads to improved separation of finer particles. In the application, a distinction is made between two possible types of filter aid addition. These are shown schematically in Figure 1.

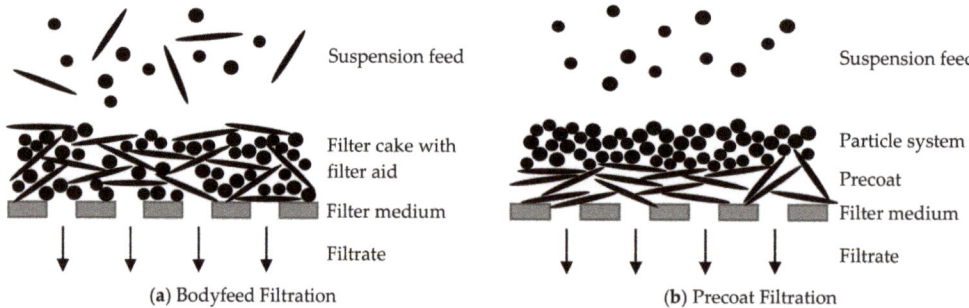

Figure 1. Continuous addition of the filter aid with the suspension (**a**) and addition as a primary filter cake (**b**).

In bodyfeed filtration, the filter aid is added to the suspension stream upstream of the filter. Here, the particle system has time to interact with the filter aid. The filter aid usually has a structure with a large surface-to-volume ratio where the particles of the suspension can adhere to it. The mixed flow of the filter aid and particle system now reaches the filter and ideally forms a lower specific cake resistance than the pure particle system. Studies on particle mixtures have shown that the specific cake resistance of fine particle systems can be reduced in this way [13]. In addition, fewer to no particles should pass through the mesh in the form of turbidity. The hydraulic load of the filtration and the quality of the filtrate are thus increased. The second application concept is regular precoat filtration. For this, a filter cake of precoat material is first built up on the fabric of the filter. In terms of equipment, a separate feed tank is required for this. The suspension then flows through this layer of precoat and the particles contained within it separate on and in the precoat cake. The turbidity shock at the beginning of filtration should be prevented by this process. In the context of this study, the focus is on filtration using precoat layers.

Precoat filtration, also known as precoat filtration, corresponds to simple cake filtration. Here, the applied cake of the precoat material serves as a filter medium for the subsequent cake filtration. Compared to a simple filter with pure surface filtration, a cake also has a depth filtration effect [14]. This also enables filtration of the finest particles down to 0.1 µm [15,16]. The filter fabric serves only as a base for the cake and can have correspondingly larger mesh sizes [1]. The main difference with simple cake filtration lies in the composition of the filter cake, which consists of several components. The composition consequently changes with continuous filtration time, as more and more components of the suspension adhere to the cake because of penetration. The precoat filter cake should be as open-pored as possible to prevent clogging of the filter cake and to keep the pressure loss across the cake low. Often, filter aids are added to the suspension to be filtered as body feed, which also makes the further cake structure as open-pored as possible, which also has a positive effect on cleaning [17,18].

In most cases, the precoat filter cake consists of a primary layer and a secondary layer. The task of the primary layer is to bridge the large pores of the filter fabric and guarantee the load-bearing capacity of the cake. It is important that the permeability remains as low as possible, since the primary layer does not have an actual filtration effect but leads to an increase in the initial resistance through interaction with the filter fabric [6]. Usually, this consists of a coarse material with a diameter larger than the pore size. The second layer contains a much larger proportion of finer particles, creating a distinct pore system responsible for the actual filtration task [12].

The selection and quantity of additives are mostly based on experience and are therefore mostly not the optimum. This is difficult to predict because it is based on complex systems and includes many variables such as material, particle-size-distribution (PSD), and concentration. In this context, many studies have investigated the amount of body feed to produce the most optimal filter cake with an open-pore structure. In this context, initial studies presented models of the dependence of the filter cake resistance on the filter aid and assumed that solid–liquid separation occurs only at the surface of the cake, which means that an open-pored cake would not be necessary at all. More recent research, however, shows the emphasized role of the depth effect of the filter cake. Here, the importance of a detailed characterization of the precoat layer to improve the filtration performance becomes clear [17].

3. Methodology

The methodical procedure of this elaboration is essentially based on three test series:

- Pressure groove tests according to [7] to determine the specific cake resistance and filter medium resistance;
- Gravimetric turbidity measurement by filtration of the filtrate from the pressure groove using a membrane;
- Backwash system according to [11] for filtration and regeneration (cake discharge).

As a standard experimental device for determining filtration characteristics, the pressurized filter cell is an essential tool for determining the specific cake resistance of the particle systems and filter aids, as well as the filter medium resistance. In all experiments, a concentration of $0.05~\text{kg}\cdot\text{L}^{-1}$ and a filtrate volume of 200 mL are used, with an effective filtration pressure of 1 bar. According to [16], the resulting specific precoat quantity of $2.5~\text{kg}\cdot\text{m}^{-2}$ ensures sufficient bridging in the precoat. Furthermore, in all tests of the pressure groove, the fabric is used only once to create the same starting conditions in all tests. Therefore, aging phenomena do not occur. The filtrate produced by the pressurized filter cell is subjected to further filtration through a membrane with a pore diameter of

0.4 µm, which is measured gravimetrically beforehand. A further determination of the membrane weight following the filtration conveys the mass of residual particles in the filtrate after the filter mesh. This procedure makes it possible to characterize the filter effect of the fabric and filter aid and to make statements about the quality of the filtration and follows a procedure with reference to [7].

Following [11], filtration is also carried out on a backflush filter system, which is illustrated in Figure 2. At the beginning of the experiment, two storage tanks are to be set up. One container is for the particle system to be filtered, while the second contains the filter aid. In both cases, care must be taken to ensure enough dispersion and possible swelling in the fluid. Following the sample preparation, the test is carried out. For this purpose, the precoat material is first fed into the process chamber and a defined quantity is built up on the filter fabric (I). The cake thickness is mainly controlled by the amount of filtrate and the concentration in the receiver tank. After a defined filtrate volume has been reached, displacement of the suspension of filter medium by clear water occurs (II). This step enables the filter cake thickness to be compared with the concentration of the filter aid in the receiver tank and the amount of filtrate that has passed through. In addition, this allows a clear cut between the precoat and particle filter cake. Once this is done, filtration starts again, but now with the particle system (I). In this case, too, a defined filtrate volume and filtrate concentration control the desired cake thickness. Subsequently, the cake is again displaced by clear water (II) to optically determine the cake thickness. By comparing both images, it is possible to define the precoat layer and particle layer. After complete displacement of the particle suspension, the filter is regenerated by backwashing with cake discharge (III). For this purpose, the backwashing fluid (deionized water) flows through the filter fabric against the direction of the filtrate, dissolves the built-up filter cake, and allows it to slide off. The cake then consolidates at the bottom of the vessel and can be discharged before the test is repeated (IV).

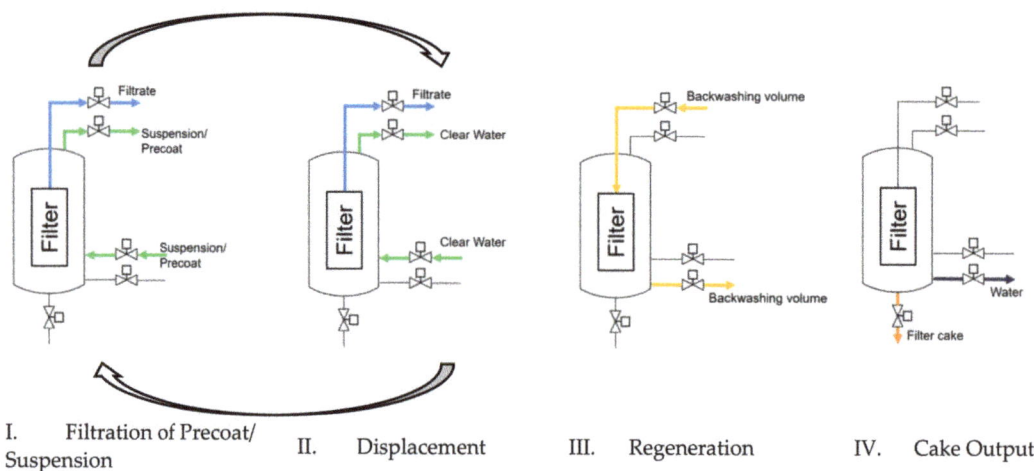

Figure 2. Schematic representation of the test plant during filtration of the precoat layer (**I**), displacement of the suspension with clear water (**II**), and subsequent running through of the same sequence with filtration of the suspension (**I**) and renewed displacement by clear water. This is followed by regeneration (**III**) through discharge of the filter cake and cake removal out of the process room (**IV**).

4. Materials

Based on the experimental procedure described in the "Methodology", the experimental matrix is carried out using the particle systems of [6,7] with the precoat material shown in Figure 3 and a satin fabric with a mesh size of 11 µm (Sefar AG, Heiden, Switzerland).

A mesh size of 11 µm is chosen because this corresponds to the mean diameter $x_{50,3}$ of the particles. This ensures that one part of the particles is rejected on surface and the other part is to be separated via bridging. Due to the scope, reference is made to the publications just mentioned for a description of the particle systems. The filter aids are described in the following section. Three precoat materials Arbocel-NV00, Filtracel and Vitacel (J. Rettenmaier & Söhne GmbH + Co KG, Rosenberg, Germany), which are shown in Figure 3, are the subject of this study.

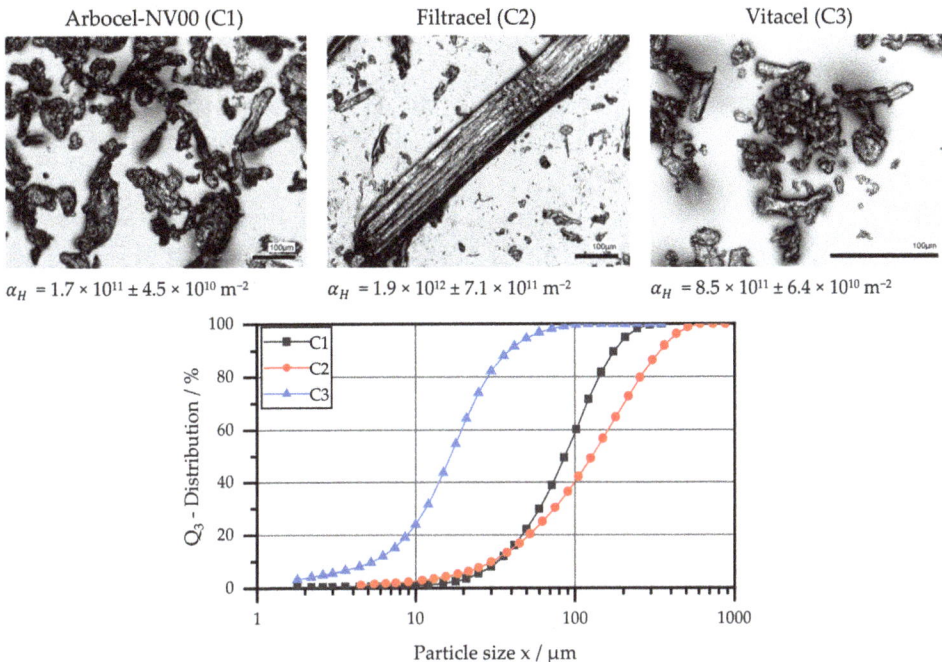

Figure 3. Microscopic image of the investigated filter aids (Cellulose) of J. Rettenmaier & Söhne GmbH + Co KG with fibrous (C1,C2) and orthorhombic (C3) structures, as well as the filter cake resistance and particle size distribution.

Precoat C1, with the manufacturer's designation Arbocel-NV00, it is a fibrous filter aid made of cellulose. A laser diffraction measurement yields a value of 87.0 µm as the average fibre length. Filtration tests with the pressure groove have shown a specific cake resistance of 1.7×10^{11} m^{-2}. Related to the mean fibre length with power 10^{11}, the Carman-Kozeny order of magnitude is plausible.

The next filter aid is Filtracel (C2) with a mean particle diameter of $x_{50,3}$ = 127.9 µm. The measured cake resistivity is 1.9×10^{12} m^{-2} and is larger than that of C1 and C3. This behavior is contrary to expectations, since cellulose C2 is the coarser filter aid after laser diffraction and thus should yield a more porous filter cake. However, Figure 3 also shows a higher particle size distribution for Precoat C2, which may be a reason for the more compact cake.

The last Precoat material, Vitacel (C3), represents the highest clarity in the food industry. This can also be seen from the particle size in Figure 3, as this cellulose is very fine, short fibres with a mean diameter of 16.7 µm. Regarding the filtration properties, this filter aid has a specific cake resistance of 8.5×10^{11} m^{-2}. This value is analogous to C1 below C2, which means that a different cake forming behavior can be assumed. All three filter aids undergo the experimental procedure described in the "Methodology" to validate the filtration and regeneration properties of the filter. This is described in the following chapter.

5. Interpretation

In this chapter, the test evaluation and interpretation will now take place based on the test materials presented in the chapter "Material" and the test procedure described in the chapter "Methodology". For this purpose, the tests are divided up and described separately based on the results from the pressure groove and backflush filtration. The following chapter, "Conclusion", contains a summary with an outlook.

5.1. Influence of a Precoat Layer on the Filter Resistance

Within the scope of the filtration tests, the filter medium resistances and the filter cake resistances during filtration with filter aids are to be investigated. For this purpose, the filter aids C1, C2, and C3 presented in the "Material" and particle systems made of [6] are each filtered and measured with the 11 µm satin fabric in the pressure groove. This is followed by the determination of the resistances for the combinations of all filter aids and the particle systems. First, a precoat layer is washed onto the filter cloth and then the suspension with the particles to be filtered is passed over the precoat.

The filter mean resistances of the resulting series of measurements are shown in Figure 4. The use of C1 results in the lowest filter mean resistances overall. C2 with the longest fibre length produces resistances with the highest values. This could be explained with the long fibres which have a higher possibility for deformation. In addition, C2 has a wider particle size distribution, making the filter cake more compact. C3, with the smallest fibre length, has a correspondingly low cake porosity due to its small average diameter, which increases the resistance. Therefore, the particle size of cellulose is less important compared to the size distribution and deformability of the cellulose fibres in this range of 17 µm to 128 µm. For the particle systems, all three are in the same order of magnitude and can be considered identical with respect to the deviation. The difference from the values published in [6] is due to the use of new tissues in each of the experiments reported here. This shows that particles are deposited differently in the tissue. Since new fabrics were always used here, there were no irreversible particle inclusions that affect the resistance. As a result, the resistance depends only on the particle size and not on the shape. The difference in shape only causes particles to embed themselves differently in the fabric and thus change the resistance during continuous tests. When several layers are combined, i.e., a precoat layer and a particle layer, the precoat represents the new filter medium. Depending on the precoat, the respective filter cake or new filter medium has a different porosity. The particle system P1 shows the clearest interaction, since the values of the filter medium resistance increase the most here. In contrast, almost no difference can be seen between P2 and P3. P1 thus penetrates much further into the filter cake and significantly increases the resistance. The orthorhombic shape consequently favors penetration into the tissue, while flaky and acicular particles are more likely to be deposited on the surface.

In addition to the filter medium resistances, the filter cake resistances show a resistance value independent of the precoat presented in Figure 5. While with the precoat the filter cake resistances still show differences in the range of 10^3, when the deviation is considered, no difference can be seen in the resistances after the particle layer has been applied. The particle systems thus dominate the resistance value, which means that there is practically no dependence of the filter cake resistance on the cellulose type. Overall, however, the filter cake resistance has decreased compared to that of the pure particle systems, which is positive for the pressure drop.

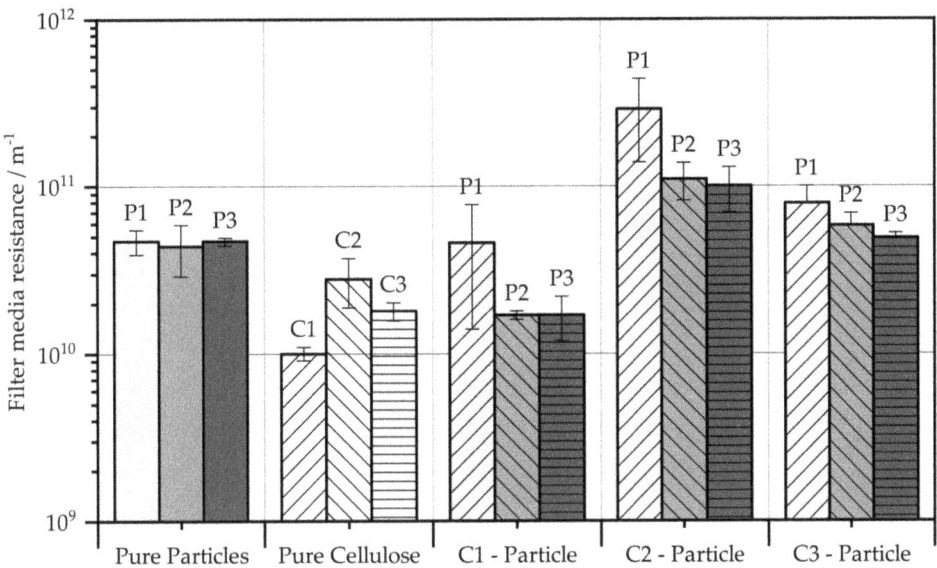

Figure 4. Filter media resistances of the pure filter aids, particle systems, and their combinations.

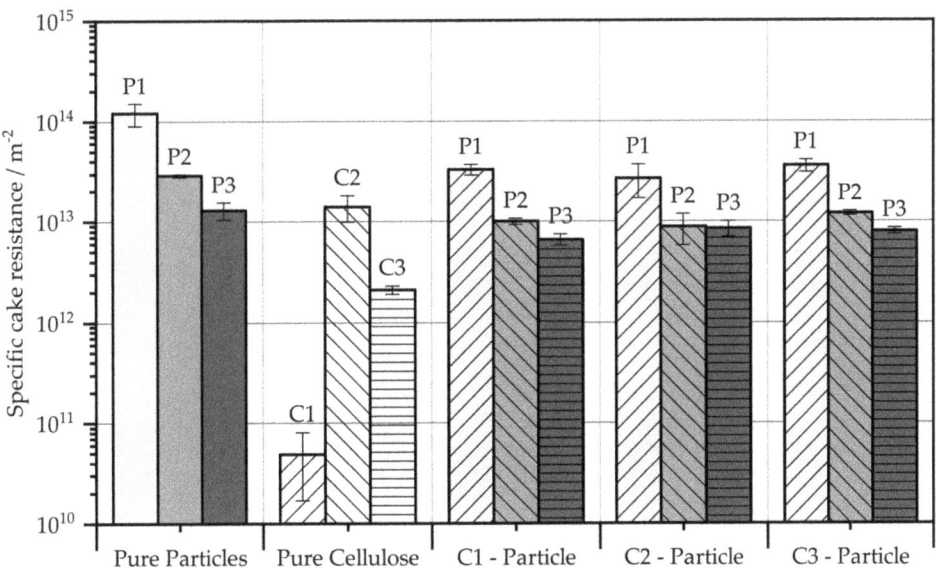

Figure 5. Filter cake resistances of the pure filter aids, particle systems, and their combinations.

5.2. Influence of Filter Aids on the Turbidity Impact

Another possibility for validating a precoat layer is the turbidity impact that occurs. As a rule, this occurs at the beginning of a filtration process until enough solid bridges have formed on the mesh. These solid bridges prevent further solid passage through the filter fabric. When a precoat layer is used, these solid bridges are already formed, reducing solid passage.

The turbidity impact was determined by the individual particle systems, filter aid, and the combination of all particles with the filter aid. Figure 6 shows the mass of residual particles per filtrate volume of all cellulose fibres used as a pure stock and precoat material. A comparison of the cellulose fibres shows a decreasing trend from C1 to C3. C1 produces the highest turbidity of 8.3 mg L^{-1} compared to the other cellulose fibres, whereas this value is halved for C2 and is only 1.3 mg L^{-1} for C3. The number of particles passing through the filter is thus a factor of 6.5 higher with cellulose C1 than when C3 is used. This is not as expected since C3 has a much smaller mean diameter than C1. One explanation is that due to the higher filter cake resistance of C3, i.e., 8.2 × 10^{11} m^{-2} compared to C1 with 1.7 × 10^{11} m^{-2}, the smaller pore size within the cake retains more particles. C2, at 7.2 × 10^{11} m^{-2}, lies between the other two fibres and confirms the trend due to the turbidity that occurs.

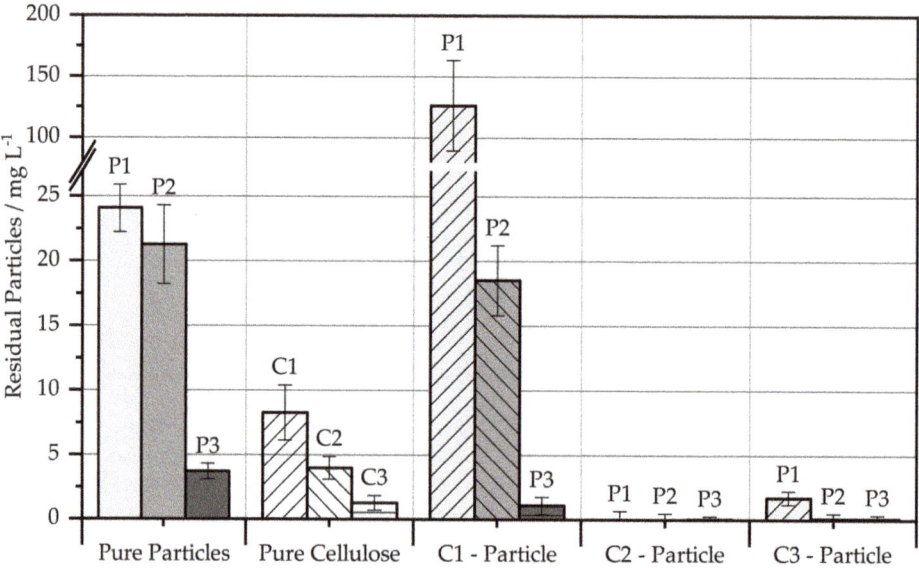

Figure 6. Mass of residual particles of the 200 mL filtrate on the pressurized filter cell using all particles and cellulose used as a pure substance and precoat.

Besides the cellulose fibres, the turbidity due to filtration of the particulate systems is higher. The peak occurs with the orthorhombic P1, i.e., 24.1 mg L^{-1}, and is the same order of magnitude with the acicular P2. Only P3, which is acicular, is characterized by a low turbidity impact of 3.7 mg L^{-1}. Surface deposition according to [19] can be seen here. The long fibres settled on the filter and prevent further penetration of particles. This explains the comparable values between the fibrous particle system P3 and the cellulose fibres.

The turbidity impact when using a precoat layer is not as expected. The turbidity of P1 when C1 is used as precoat is striking. At 126.3 mg L^{-1}, this is a factor of five above the turbidity of the pure particle system P1, which at first sight seems illogical. The reason lies in the pore size of the filter cake. This increases with increasing fibre length. Small particles of P1 thus do not deposit on the cake surface and penetrate the cake. Particles are deposited in the cake due to a depth effect and lower the particle concentration. Particles that have passed through the cake have subsequently decreased in concentration to a level through which bridging occurs later [20]. Thus, more particles can penetrate through the tissue. In P2 with Precoat C1, the turbidity impact of 18.5 mg L^{-1} is lower than the pure particle system, but still the same order of magnitude with respect to the deviation. Thus, no marked decrease of the solid content in the filtrate can be seen in this combination either. Only P3 shows a smaller turbidity impact of 1.1 mg L^{-1}, which can be attributed to the shape of the particles. The particulate and cellulose fibres interlock and thus prevent the passage of solids. With fibre C2, contrary to expectation, the turbidity shock for all particle systems is at the limit of determination of the balance used and is thus practically non-existent. This can be explained by the greater swelling behaviour of C2 in water. Together with the higher deformability of longer fibres, a compressible layer is created close to the tissue through which the particles cannot penetrate. This is often called the skin effect [21]. C3 also shows a clear decrease in turbidity. This is the only cellulose fibre that exhibits surface filtration and deposits most particles at the surface. Only P1 shows a measurable solid content of 1.7 mg L^{-1} with C3, although this is still 14 times smaller than that with the pure stock.

As an overall result, C2 as a precoat material has the lowest number of solids in the filtrate, with C3 providing similar values. C1, on the other hand, is unsuitable as a precoat material regarding solid passage since it increases the turbidity impact at P1. The difference in turbidity impact between celluloses C1, C2, and C3 also confirms the importance of a depth effect in cake filtration according to [14].

5.3. Influence of a Precoat Layer on the Backwashing Behavior

Sufficient validation requires the backwashing of the individual particle systems, as well as the individual cellulose fibres in detail. Subsequently, a fixed cake height of the precoat layer for filtration of the particle systems must be determined to make them comparable to each other. There are several options for determining the cake height of the precoat layer.

Option 1

One possibility is the self-determination of a suitable cake height, which is based on the complete discharge [11]. This procedure has the advantage that the cake height is determined with the same basic conditions as in the subsequent use as a precoat. Thus, errors due to environmental conditions, such as the composition of the water for the suspension, which affects the agglomeration behavior and viscosity, can be avoided in the cake height determination. If a complete discharge is not possible, the cake heights are based on the cellulose amounts of the respective cellulose fibres that produce a complete regeneration. It is impossible to generate a complete discharge even with pure cellulose; the layer thickness should be about 0.5–2 mm so that the complete surface is covered. The fabric used for filtration has a mesh size of 11 µm and a satin weave with the designation 30-04 01-01-02 according to DIN 9354 (Deutsches Institut für Normung).

Option 2

Based on the manufacturer's specification, filtration companies further develop their products and can draw on many years of experience. A quantity of 1 kg·m^{-2} was suggested as the manufacturer's specification for the layer thickness of the cellulose. The filter cartridge used has a diameter of 31 mm and a height of 110 mm, which results in a filtration area of 10^7 cm^2. Thus, 10.7 g of cellulose should be used regardless of the cellulose type. The cake heights are shown in Figure 9.

In the context of this paper, both options are used for their suitability for evaluating the backwashing properties of the precoat layers and particle-laden precoat layers. During regeneration, a backflush pressure of 0.5 bar is used, which directs clean water through the fabric against the direction of filtrate. The termination criterion is a 5 L·m^{-2} backwashing volume. This quantity is more than double the guideline of the 2 L·m^{-2} backwash volume given by [22] in the industry. Thus, sufficient validation of the filter cake discharge can be carried out. If the filter cake is not completely removed from the fabric afterwards, the regeneration counts as incomplete. Subsequently, the concentration, or filtrate quantity, increases, which results in a greater cake height. This provides more cohesive forces in the cake, causing it to drop in larger segments, allowing for more complete cleaning. The discharge to be achieved is the complete discharge, which represents the optimum for the required time and fluid quantity for backflush filtration. In the following subchapters, the backwashing of the individual particle systems, cellulose fibres, and their combination will be presented and defined.

5.3.1. Cleaning Behavior of the Pure Particle Systems

Analogous to [6], the pure particle systems consist of SF300, Tremin283-100, and Tremin939-304 from Quarzwerke GmbH. These have similar particle sizes of $x_{50,3} < 25$ µm, but different structures. SF300 (P1) has an orthorhombic particle structure, Tremin283-100 (P2) shows a flaky structure, and Tremin939-304 (P3) is needle-shaped. Regeneration of the pure particle systems shows a decrease in the required cake height from P1 to P2 to P3. According to [23], a complete discharge occurs only when the flow force of the backwash fluid is greater than the adhesion force but not greater than the cohesion force. If this statement is used to explain it, either a decrease in the cohesive forces in the filter cake from P3 to P1 or different adhesion forces between cake and mesh hold true for a constant cake discharge over the complete series of experiments. The needle-shaped P3, which is made of the same material as the plate-shaped P2, shows a threefold decrease in the required cake height for complete regeneration. In this context, tests with the pressure groove show that the filter cake of P3 is much more open-pored than that of the two particle systems P1 and P2. Since, according to Karman-Cozeny, the cohesive forces in the cake behave according to the porosity; they should therefore be greater for P1 and P2 than for P3, which would require a lower cake height. The adhesion force, on the other hand, depends on the particle interaction with the tissue. Thus, it follows that the adhesion forces must be greater for P1 and P2 than when P3 is used. Here, the dependence of the particle shape becomes apparent. P3, with its acicular structure, exhibits surface deposition, resulting in little interaction with the tissue, and a low cake height is sufficient for discharge. With particles P2 and P3, the adhesion force is greater, as a much greater cake height is required for complete cleaning, shown in Table 1. This indicates a deeper penetration of particles into the tissue, whereby bleeding, or clogging effects must be assumed. Another striking feature is the increased backwashing volume at P1. This is due to a partial discharge of the cake, which can be attributed to insufficient cohesive forces or excessive adhesive forces. The idea here is to use a precoat layer to reduce the adhesion forces so that the interaction of the particle systems with the fabric is reduced.

Table 1. Regeneration behavior in a satin fabric with a mesh size of 11 µm and particle systems based on [8,11].

Particle	Concentration/kg·L^{-1}	Particle Volume/kg·m^{-2}	Cake Height/mm	Backwashing Volume/L·m^{-2}	Remark
P1	5.71	1.07	0.7 ± 0.11	2.2 ± 0.50	Complete Discharge
P2	5.71	1.07	0.6 ± 0.07	0.3 ± 0.01	
P3	1.43	0.13	0.2 ± 0.07	0.3 ± 0.03	

5.3.2. Cleaning Behavior of Pure Cellulose

For a sufficient validation of the cake discharge, the release behavior of the pure cellulose layer still has to be characterized. For this purpose, filtration with subsequent cake release is carried out analogously to Figure 2 with variable cake thickness. The aim is to determine the complete discharge for the variation according to option 1.

For pure cellulose, C1 does not generate a complete discharge, since the cohesive forces are too low to hold the filter cake together during backwashing. Thus, the filter cake immediately disperses again during the backwashing process; see Figure 7.

(a) $t_0 = 0$ s   (b) $t_1 = 1.2$ s   (c) $t_2 = 2.16$ s

Figure 7. Backwashing behavior of the pure C1 layer with a surface loading of 0.13 kg·m^{-2} and the resulting cake height of 0.7 mm.

Here, it would be possible to obtain a higher compression of the filter cake via a higher filtration pressure. Compression or rearrangement in the filter cake results in a smaller pore size of the pile and thus increases the cohesive forces between the particles. A mechanically more stable filter cake is the result. By means of this increasing stability, direct dispersion after detachment of the cake from the mesh by the backwash process should be prevented. The aim is to generate a sheath release even with needle-shaped cellulose particles. Individual tests have shown that compaction with pressures up to 2 bar do not achieve the desired effect of further compaction. This filter cake is comparable to the cakes at 1 bar in diameter and does not bring any improvement in the cleaning behavior, which is why no further tests will be carried out here.

A striking feature of the cake formation is the formation of spherical cellulose heaps on the fabric. During filtration, these form "mountains and valleys", which lead to an uneven cake diameter; see Figure 8. One reason for this is the low specific cake resistance. This means that the particles do not accumulate in the thinnest parts of the cake and compensate for differences in height. With small cake resistances, the total resistance of the cake and fabric increases only slightly with increasing cake height. However, the flow is reciprocal to the resistance. If the resistance is higher at any point, less fluid also flows through it and the cake formation rate decreases. Thus, a uniform cake is formed for particles with greater cake resistance [11].

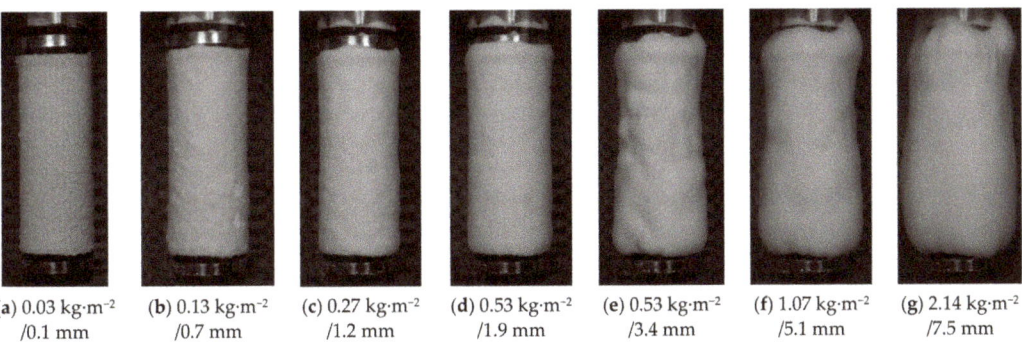

(a) 0.03 kg·m^{-2} /0.1 mm   (b) 0.13 kg·m^{-2} /0.7 mm   (c) 0.27 kg·m^{-2} /1.2 mm   (d) 0.53 kg·m^{-2} /1.9 mm   (e) 0.53 kg·m^{-2} /3.4 mm   (f) 1.07 kg·m^{-2} /5.1 mm   (g) 2.14 kg·m^{-2} /7.5 mm

Figure 8. Resultant cake thickness at time t_0 with ascending area occupancy.

C1 has a small cake resistance of 4.5×10^{11} m^{-2} in this fabric, which increases the total resistance only insignificantly. In comparison, the cake resistance of P1 is 4.2×10^{13} m^{-2} and is larger than cellulose by a factor of 10^2. Accordingly, the cake resistance is the determining factor, which indicates the flow velocity through the filter. This results in very non-uniform cakes, since the flow resistance of the forming cake is relatively small and there is no flow-related equalization of the cake thickness. A disadvantage is the non-uniform distribution of the particles with respect to the subsequent filtration with cellulose as a precoat layer. It is to be expected that with low precoat thickness, part of the filter will not be occupied and thus will not be able to perform its purpose as a filter aid at this point; see Figure 8a). By the time the individual cellulose particles bind together to form a fully wetted surface, the cake has a height of between 0.6 and 0.8 mm. A multiple determination with a constant cellulose quantity of 1.43 g then gives a cake thickness of 0.7 mm, which is why this size is used for the following precoating.

In Figure 8, the difference between the two cakes with a surface occupancy of 0.53 kg·m^{-2} lies in an increasing suspension concentration for e. Here, the influence of the suspension concentration can be seen, analogous to [20]. The cake heights used for the following precoat filtration are shown in Figure 9.

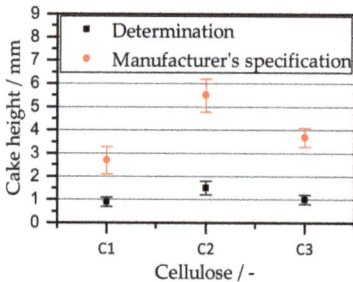

Figure 9. Cake height of the precoat layer.

5.3.3. Cleaning Behavior with Additional Precoat Layer

Several differences can be identified between the backwashes of the pure particle systems and the backwashes with the precoat layer.

In filtration with an additional precoat layer, the basic shape of the filter cake depends on the precoat. The particle system, which usually has a much higher specific cake resistance, then lies evenly on the precoat layer. Unevenness in the precoat means an uneven filter cake, as can be seen from Figure 8d,e. Here, as a result of the high permeability, the precoat layer has accumulated in the lower half of the filter cartridge, creating a drop shape on which the filter cake subsequently builds up. This is counterproductive for a complete discharge, since there is always a predetermined breaking point at the thinnest part of the cake in the case of unevenness.

Another effect of precoating is that the overall diameter increases with an additional precoat layer. With constant concentration and filtrate quantity of the particle system, this consequently leads to a thinner filter cake, which has lower stability. An evaluation of LSM images of the filter cake on the pressurized filter cell has shown that particles not only settle on the surface of the cake, but also penetrate the cake and deposit there via depth filtration of the filter cake. Accordingly, an exact determination of the height of the individual layers is not possible in this combination of C1 and P1. This can also be explained by the PGV of C1 and P1. The $x_{90,3}$ value of P1 with 28.95 μm is below the $x_{10,3}$ value of C1 with 32.95 μm. This means that the largest particles in P1 are still several μm smaller than the smallest of cellulose C1. For the particles to be deposited on the surface, the quotient of the pore size of the cake particle size must not fall below 0.4 to achieve a sieving effect or to allow solid bridges [19]. The relationship between pore size and particle diameter is represented by the Young–Laplace equation for a heap [24]. From this equation, assuming a monomodal system, a pore size of 2/3 of the Sauter diameter can be determined. This explains the particle breakdown of P1 through the filter cake. The resulting cake diameters and backwash quantities are shown in Tables 2 and 3.

Table 2. Cake height and backwashing volume at a cellulose quantity of 0.13 kg·m^{-2} (option 1) with the respective particle systems at a backwashing pressure of 0.5 bar.

Particle	Concentration/kg·L^{-1}	Particle Volume/kg·m^{-2}	Cake Height/mm	Backwashing Volume/L·m^{-2}	Remark
C1–P1	22.86	2.13	1.4 ± 0.4	1.40 ± 0.22	Incomplete discharge
C1–P2	1.43	0.13	1.1 ± 0.2	1.19 ± 0.10	Resuspended discharge
C1–P3	1.43	0.13	1.2 ± 0.2	0.53 ± 0.06	Complete discharge
C2–P1	1.43	0.13	2.0 ± 0.3	> 5	No filtration effect
C2–P2	1.43	0.13	1.5 ± 0.3	> 5	No filtration effect
C2–P3	1.43	0.13	1.4 ± 0.4	1.55 ± 0.18	Incomplete good discharge
C3–P1	1.43	0.13	0.9 ± 0.3	0.76 ± 0.08	Complete discharge
C3–P2	1.43	0.13	0.9 ± 0.1	0.23 ± 0.02	Complete discharge
C3–P3	1.43	0.13	0.9 ± 0.2	0.24 ± 0.04	Complete discharge

Table 3. Cake height and backwashing volume at a cellulose quantity of 1 kg·m^{-2} (option 2) with particle system P1 at a backwashing pressure of 0.5 bar.

Particle	Concentration/kg·L^{-1}	Particle Volume/kg·m^{-2}	Cake Height/mm	Backwashing Volume/L·m^{-2}	Remark
C1–P1	22.86	2.13	3.1 ± 0.2	1.07 ± 0.10	Resuspended discharge
C2–P1	5.71	1.07	5.3 ± 0.7	>5	No filtration effect
C3–P1	1.43	0.13	2.9 ± 0.3	2.19 ± 0.14	Complete discharge

This shows that when C1 is used, a larger backflush volume is required because of an increasing backwashing duration at a constant backwashing volume flow. For a total cake thickness of 1.4 mm, this amounts to 1.4 L·m^{-2}. Compared to the complete discharge of the pure particle system with a required backwashing volume of 0.23 L·m^{-2}, this is an additional volume requirement by a factor of six. Cellulose C1 is thus considered unsuitable not only for turbidity measurement but also with a focus on regeneration quality.

C2 has the largest particles when viewed from the PGV and therefore forms a cake with the largest pores. This is evident during filtration, as no particles from P1 and P2 are deposited on the cake surface. The particles deposit in the cake, with the main part collecting directly on the tissue surface. The fact that there is nevertheless no particle breakdown after turbidity measurement is attributed to the swelling behavior and compression close to the filter, named the skin effect. During the subsequent regeneration, only the precoat layer comes off and the particle layer remains adhered to the filter. Increasing the cake thickness to that of the manufacturer according to option 2 also does not improve the regeneration. Only the particle system P3 shows surface filtration with C2 as the precoat and is completely removed even after cleaning. However, clean regeneration is possible with P3 even without the Precoat, which makes the use of Precoat unnecessary. Therefore, the cellulose fibre C2 is classified as "unsuitable".

Of all the cellulose fibres tested, C3 is the only one that can produce a complete discharge at 1.0 mm cake thickness. The filtration of the particle systems is now carried out for this cake height at different heights with subsequent backwashing. At all measured cake heights of the particle system, a complete discharge can be seen. Accordingly, the particles have no direct influence on the cleaning behavior. A similar behavior can be observed at the application rate recommended by the manufacturer. Here, too, the discharge of the filter cake behaves similarly to the discharge of the pure precoat layer. The manufacturer's suggested quantity of 1 kg·m^{-2} is therefore too high, and according to the tests, a quantity of 0.13 kg·m^{-2} cellulose is sufficient for a cleansing with optically complete regeneration, independent of the particle quantity; see Figure 10a–c.

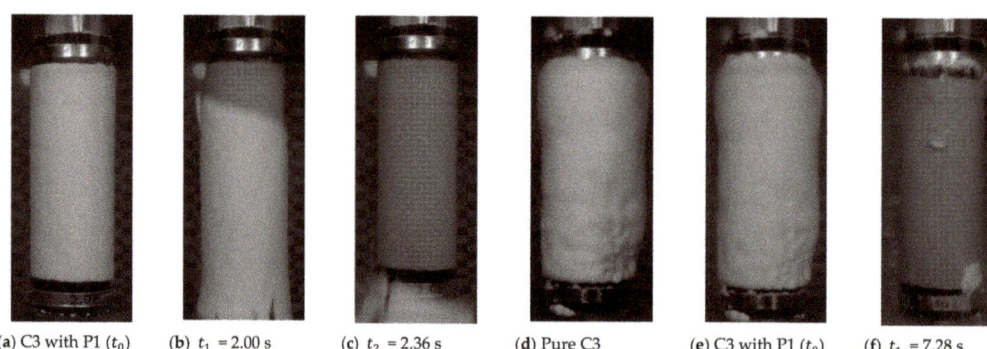

(a) C3 with P1 (t_0) (b) t_1 = 2.00 s (c) t_2 = 2.36 s (d) Pure C3 (e) C3 with P1 (t_0) (f) t_1 = 7.28 s

Figure 10. Regeneration of C1 and P1 with 0 13 kg·m^{-2} each for a new filter (**a–c**) and a regeneration of C1 (1.07 kg·m^{-2}) with P1 (2.13 kg·m^{-2}) for a dirty filter (**d–f**).

When using a used filter, an uneven filter cake is also present, similar to C1. If the filter is extremely dirty, there may be areas without continuous precoat wetting; see Figure 10d–f. On subsequent backwashing, this leaves residual contamination on the filter in zones without a precoat. Such contamination of the filter cannot be removed by backwashing and must be treated by other cleaning methods. These include cleaning by acids or ultrasonic methods [25]. The problem with these sites is that this area can no longer be used for filtration. The free filter area and thus the possible hydraulic load of the filter therefore decrease. Consequently, if a fabric is contaminated, an additional application of a precoat layer does not help to reduce the surface loading of the filter after regeneration.

6. Conclusions

The experiments carried out within the scope of this investigation include the influence of the filter aid as a precoat on:

- the filter medium and filter cake resistance according to [7];
- the gravimetrically determined turbidity impact;
- the cake release at the backwash filter according to [8].

From the pressurized filter cell tests, it was observed that the difference in shape of the particles is decisive for the depth of penetration into the fabric. Needle and flaky shaped particles tend to deposit on the surface of the fabric, whereas particles with an orthorhombic structure can penetrate far into the fabric and increase the filter media resistance in continuous tests. No dependence on the precoat material can be seen in the filter cake resistance since this usually has a much lower resistance to the particle system. The filter cake resistance is therefore dominated by the resistance of the particle system but is constantly below that of pure particle systems due to slight mixing with the filter aid.

The advantage of precoating with a lower turbidity impact cannot be demonstrated with all cellulose fibres. While with C2 and C3, the turbidity impact is practically eliminated, with C1, five times the amount of particles penetrated through the fabric. Therefore, no general improvement of the turbidity impact can be assumed through the use of filter aids. In general, the filter aid resistance is a good method of determining the suitability of the filter aids since the turbidity impact behaves according to this.

On the backwashing filter, an improvement in regeneration performance can be noted with C3 as the filter aid. With C3, the complete discharge is independent of the number of particles filtered by the particle system. Thus, a particle layer of any thickness can be applied to the precoat layer in combination with the 11 µm satin fabric since the adhesion forces that cause the cake to adhere to the fabric depend only on the filter aid [23]. This in turn is offset by a larger required backwash volume, up to three times, because of the greater cake height. Increasing the amount of precoat does not improve the regeneration result. C1 shows a resuspension tendency during regeneration. The necessary cohesive force to hold the cake together comes only from the subsequent particle system. Thus, again, large layer thicknesses are needed to regenerate this in a complete discharge. Without a complete discharge, a large amount of fluid must subsequently be used to remove the remaining cake. Increasing the layer thickness enhances depth filtration but is not proportional to the improvement in regeneration.

Author Contributions: Conceptualization, V.B. and P.M.; Data curation, V.B.; Formal analysis, V.B. and P.M.; Funding acquisition, P.M. and H.N.; Investigation, V.B.; Methodology, V.B.; Project administration, P.M.; Resources, P.M.; Software, V.B.; Supervision, P.M., M.G. and H.N.; Validation, V.B.; Visualization, V.B.; Writing—original draft, V.B.; Writing—review & editing, P.M., M.G. and H.N. All authors have read and agreed to the published version of the manuscript.

Funding: This research received no external funding.

Institutional Review Board Statement: Not applicable.

Informed Consent Statement: Not applicable.

Data Availability Statement: Not applicable.

Acknowledgments: The authors would like to thank the German Federation of Industrial Research Associations (AiF) for the financial support (IGF number 18591 N). The authors would also like to thank all colleagues and students for the support in writing this paper. Furthermore, the authors would like to thank the company J. Rettenmaier & Söhne GmbH + Co KG, and especially Eberhard Gerdes and Sebastian Lösch, for providing and advising the selection of suitable precoat materials. We acknowledge support from the KIT-Publication Fund of the Karlsruhe Institute of Technology.

Conflicts of Interest: The authors declare no conflict of interest.

Abbreviations

Symbol	Description	Unit
A	Filter area	m^2
d_h	Hydraulic diameter	m
F_H	Adhesive force	N
H_{FC}	Filter cake height	m
R_{FM}	Filter media resistance	m^{-1}
V_F	Filtrate volume	m^3
$\dot{V}_F$	Filtrate volume flow	$m^3 \cdot s^{-1}$
$x_{50,3}$	Mass/volume modal value	m
α_H	Specific filter cake resistance	m^{-2}
Δp	Pressure difference	Pa
ε	Porosity	-
η_f	Viscosity of the fluid	$Pa \cdot s$

References

1. Purchas, D.B.; Sutherland, K. *Handbook of Filter Media*, 2nd ed.; Elsevier: Amsterdam, The Netherlands, 2002.
2. Ripperger, S.; Schnitzer, C. Die Barrierewirkung von Geweben Teil 1: Textiltechnische Charakterisierung und Barrieremechanismen. *Filtr. Sep.* **2005**, *19*, 110–117.
3. Schnitzer, C.; Ripperger, S. Die Barrierewirkung von Geweben Teil 2: Experimentelle Methoden zur Bestimmung von Gewebeeigenschaften. *Filtr. Sep.* **2005**, *19*, 166–173.
4. Ripperger, S. Die Barrierewirkung von Geweben Teil 3: Auswahl von Geweben zur Kuchenfiltration. *Filtr. Sep.* **2005**, *19*, 284–289.
5. Ripperger, S. Die Barrierewirkung von Geweben Teil 6: Einfluss des Filtermittels auf die Kuchenfiltration. *Filtr. Sep.* **2008**, *22*, 6–9.
6. Bächle, V.; Morsch, P.; Fränkle, B.; Gleiß, M.; Nirschl, H. Interaction of Particles and Filter Fabric in Ultrafine Filtration. *Eng. Adv. Eng.* **2021**, *2*, 9. [CrossRef]

7. Verein Deutscher Ingenieure. *VDI 2762: Filtrierbarkeit von Suspensionen Bestimmung des Filterkuchenwiderstands*; Verein Deutscher Ingenieure e.V.: Düsseldorf, Germany, 2010.
8. Morsch, P.; Anlauf, H.; Nirschl, H. The influence of filter cloth on cake discharge performances during backwashing into liquid phase. *Sep. Purif. Technol.* **2021**, *254*, 117549. [CrossRef]
9. Fränkle, B.; Morsch, P.; Nirschl, H. Regeneration assessments of filter fabrics of filter presses in the mining sector. *Miner. Eng.* **2021**, *168*, 106922. [CrossRef]
10. Christensen, M.L.; Klausen, M.M.; Christensen, P.V. Test of precoat filtration technology for treatment of swimming pool water. *Water Sci. Technol.* **2017**, *77*, 748–758. [CrossRef]
11. Morsch, P.; Ginisty, P.; Anlauf, H.; Nirschl, H. Factors influencing backwashing operation in the liquid phase after cake filtration. *Chem. Eng. Sci.* **2020**, *213*. [CrossRef]
12. Hackl, A.; Heidenreich, E.; Hoflinger, W.; Tittel, R. Filterhilfsmittelfiltration. *Fortschr. VDI Reihe 3* **1993**, *348*, 1–5.
13. Heertjes, P.; Zuideveld, P. Clarification of liquids using filter aids Part III. Cake Resistance in surface filtration. *Powder Technol.* **1978**, *19*, 45–64. [CrossRef]
14. Heertjes, P.M.; Zuideveld, P.L. Clarification of liquids using filter aids Part II. Depth filtration. *Powder Technol.* **1978**, *19*, 31–43. [CrossRef]
15. Khirouni, N.; Charvet, A.; Drisket, C.; Ginestet, A.; Thomas, D.; Bémer, D. Precoating for improving the cleaning of filter media clogged with metallic nanoparticles. *Process. Saf. Environ. Prot.* **2021**, *147*, 311–319. [CrossRef]
16. Zeller, A. Substitution der Kieselgur Durch Regenerierbare Zellulosefasern Auf Einem Neuartigen Filtrationssystem Für Brauereien. Ph.D. Thesis, Technische Universität Bergakademie Freiberg, Freiberg, Germany, 2011.
17. Kain, J. Entwicklung und Verfahrenstechnik Eines Kerzenfiltersystems (Twin-Flow-System) als Anschwemmfilter. Ph.D. Thesis, Technische Universität München, München, Germany, 2005.
18. Kuhn, M.; Briesen, H. Dynamic Modeling of Filter-Aid Filtration Including Surface- and Depth-Filtration Effects. *Chem. Eng. Technol.* **2016**, *39*, 425–434. [CrossRef]
19. Rushton, A. Effect of Filter Cloth Structure on Flow Resistance, Bleeding, Blinding and Plant performance. *Chem. Eng.* **1970**, *273*, 88–94.
20. Anlauf, H. Filtermedien zur Kuchenfiltration-Schnittstelle Zwischen Suspension und Apparat. *Chem. Ing. Tech.* **2007**, *79*, 1821–1831. [CrossRef]
21. Tiller, F.M.; Green, T.C. Role of porosity in filtration IX skin effect with highly compressible materials. *AIChE J.* **1973**, *19*, 1266–1269. [CrossRef]
22. Ripperger, S. Optimierung von Rückspülfiltern für Flüssigkeiten. *Filtr. Sep.* **2008**, *22*, 68–72.
23. Morris, K.; Allen, R.W.; Clift, R. Adhesion of Cakes to Filter Media. *Filtr. Sep.* **1987**, *24*, 41–45.
24. Lengweiler, P. Modelling Deposition and Resuspension of Particles on and from Surfaces. Ph.D. Thesis, Swiss Federal Institute of Technology Zürich, Zürich, Switzerland, 2000.
25. Gruschwitz, F.; Nirschl, H.; Anlauf, H. Optimized backflushing process for fibrous media in engine oil filtration and enhancement by ultrasound. *Chem. Eng. Technol.* **2013**, *36*, 467–473. [CrossRef]

Article

Elevated LNG Vapour Dispersion—Effects of Topography, Obstruction and Phase Change

Felicia Tan [1], Vincent H. Y. Tam [2,*] and Chris Savvides [1,†]

[1] BP, I&E Engineering, Sunbury on Thames, London TW16 7LN, UK; Felicia.Tan@uk.bp.com
[2] School of Engineering, Warwick University, Coventry CV4 7AL, UK
* Correspondence: v.tam@warwick.ac.uk
† Retired, formerly BP, no email address.

Abstract: The dispersion of vapour of liquefied natural gas (LNG) is generally assumed to be from a liquid spill on the ground in hazard and risk analysis. However, this cold vapour could be discharged at height through cold venting. While there is similarity to the situation where a heavier-than-air gas, e.g., CO_2, is discharged through tall vent stacks, LNG vapour is cold and induces phase change of ambient moisture leading to changes in the thermodynamics as the vapour disperses. A recent unplanned cold venting of LNG vapour event due to failure of a pilot, provided valuable data for further analysis. This event was studied using CFD under steady-state conditions and incorporating the effect of thermodynamics due to phase change of atmospheric moisture. As the vast majority of processing plants do not reside on flat planes, the effect of surrounding topography was also investigated. This case study highlighted that integral dispersion model was not applicable as key assumptions used to derive the models were violated and suggested guidance and methodologies appropriate for modelling cold vent and flame out situations for elevated vents.

Keywords: LNG vapour; dispersion; hazard distances; CFD; topography; phase change; cold venting

Citation: Tan, F.; Tam, V.H.Y.; Savvides, C. Elevated LNG Vapour Dispersion—Effects of Topography, Obstruction and Phase Change. *Eng* **2021**, *2*, 249–266. https://doi.org/10.3390/eng2020016

Academic Editor: Antonio Gil Bravo

Received: 6 April 2021
Accepted: 11 June 2021
Published: 15 June 2021

Publisher's Note: MDPI stays neutral with regard to jurisdictional claims in published maps and institutional affiliations.

Copyright: © 2021 by the authors. Licensee MDPI, Basel, Switzerland. This article is an open access article distributed under the terms and conditions of the Creative Commons Attribution (CC BY) license (https://creativecommons.org/licenses/by/4.0/).

1. Introduction

It is common to use the integral jet dispersion model to assess hazard distances of the discharge of flammable gases for elevated discharges, such as a tall vent stack or from an elevated flare during cold venting. The use of the integral jet dispersion model assumes that the discharge occurs in uniform ambient wind field with no orographic effect and there are no heat sources or sinks involved in the thermodynamics in the dispersion processes. In many real situations, these assumptions are not valid. In this paper, a case study is described, which illustrates situations where these assumptions are not valid and the consequences of the outcome of assessment.

This paper illustrates the concerns of dispersion of a very cold momentum gas jet from height. The scenario involved the discharge of liquefied natural gas (LNG) vapour at speed and at a height through an elevated vent stack. One of the purposes of a vent stack is for the safe dispersal of routine or emergency release of flammable gases. Some vent stacks have a pilot light to allow the flaring of these gases. In the event of the failure or absence of a pilot light, stack design ensures that the emitted gas would not be hazardous to people or facilities downwind. By virtue of its height and location, vent or flare stacks, in general, are designed to be inherently safe: disperse harmlessly in the atmosphere and not posing flammable or toxic hazard to personnel or the process facilities. This is done via a few methods such as locating the stack at a far enough distance from the facility, a high enough stack or optimal diameter of stack to allow high velocity venting. This ensures turbulence has sufficient time and intensity to dilute the discharged gas to harmless concentrations.

1.1. Common LNG Dispersion Scenarios

It is assumed generally, in hazard and risk analysis, that a spill of LNG is on the ground or on the water or sea. This covers scenarios such as a spill from low level pipework, leaks from storage tanks and spills into the sea during loading and unloading from LNG tankers. The physics involved is relatively simple to conceptualise and model.

The vast majority of consequence analysis for risk analysis purposes use evaporation-and-boiling model in conjunction with integral dense gas dispersion model for these situations. The most common model in used is dense gas dispersion model based on work by Haven [1] for spills on land/sea/water.

1.2. A Less Common Scenario

In the less common scenario, cold LNG vapour is vented in an elevated position, as mentioned earlier. This is usually addressed at the design of the vent stack/flare, the approach is to use an integral atmospheric jet dispersion model such as the one based on Ooms [2]. Integral atmospheric jet dispersion models are widely used to estimate flammable hazard distances, for design (e.g., zoning of hazard zones, in the design of cold vents to define vent heights required for safe dispersal, etc.) and for risk assessment (e.g., small to large leaks in hydrocarbon process areas). Flammable hazard distances are often defined by the distances to half of the lower flammability limits (LFL) (or sometime to LFL) of the released gas mixtures in air, and they are often calculated accordingly.

Integral atmospheric jet dispersion models are simple and can be easily solved using office computers. However, these models contain many assumptions some of which are physically unrealistic for LNG vapour dispersion assessment.

1.3. Assumptions

Current integral jet dispersion models are based on the elevated dispersion model developed in the early 1970s to assess effects of pollutants from tall stacks for environmental impact assessment. Coal fired power stations were common. Many of them were located close to or within heavily populated cities (e.g., the Battersea power station in London). The prime application of the model was the assessment of the impact of the pollutant, particularly sulphur, on air quality on communities immediate downwind of the chimney. A jet dispersion model (also called high momentum dispersion model, or elevated plume model) describes the dispersion of a gas discharged at velocities, significantly higher than ambient wind velocities.

Integral atmospheric jet dispersion models are usually embedded in commercially available consequence analysis packages such as PHAST by DNVGL for the oil/gas/petrochemical industry. Some major international companies and consultants have their own in-house package, e.g., CIRRUS in BP or FRED in Shell. The most widely used commercial packages have become the 'industry standard' and their use is sometime written into technical guidance and practices (e.g., [3]).

These integral models assume uniform conditions to simplify the physics and mathematics involved. The key assumptions are: (i) constant wind velocities, (ii) constant and uniform turbulence that implies perfectly flat terrain and no large buildings or equipment close by upstream and downstream and (iii) there is no mass or energy source or sinks, which implies there is no phase change. In practice all the above assumptions are violated for the dispersion of cold LNG vapour.

1.4. Objectives

The objectives of this analysis were to model using computational fluid dynamics (CFD) and quantify the effects of the common assumptions on conditions typically found on an LNG installation, such as topography and cold temperature of the LNG vapour, and to compare calculated results with data obtained from a recent unplanned cold venting of LNG vapour event due to failure of a pilot.

1.5. Cold Venting Incident—A Case Study

This incident occurred during the night. There was a planned maintenance that required LNG to be discharged to flare. On this occasion, the pilot failed. This led to a large amount of cold boil-off LNG vapour (estimated to be about 200 tonne) vented from a 35 m tall flare stack over a 5 h period. Cold venting can be part of planned actions. However, unintended incidents involving cold venting are often the results of the flare pilot systems failure or being isolated under specific known operational conditions. In this incident, the pilot system was inactive and the extinguished pilot alarms were left unnoticed by the facility control room operator. This incident resulted in the triggering of the facility gas detection system, specifically, line-of-sight detectors located in several locations at the processing facility to the south and downwind of the stack. The north of the facility (closest point to the stack) is situated approximately 800 m away from the flare. The gas concentration measurements in the region of interest (gas detectors that were activated at the facility during the incident were situated at the north sector of the facility) were logged and were used in this modelling analysis. The gas concentration was recorded in terms of concentration in unit of lower flammability limit integrated over distance in metre (LFL m) and it was of >3 LFL m over a distance of 13.5 m (average concentration of >0.2 LFL/m over this distance) with these values corresponded to average concentrations over a period of 30 s. The corresponding LFL reading during the incident was 2–15% LFL. It should be noted that the low set point of these line-of-sight (LOS) gas detectors was 0.2 LFL m or 20% LFL for 1 m coverage. Point detectors on the facility had set points of 20% LFL. Thus, it was possible to trigger the line-of-sight detectors without triggering point detectors, and vice versa.

2. Methodology

2.1. The Facilities

Figure 1 is a schematic layout of the LNG facility chosen for this study. The figure shows the location of the vent stack and topographical features. This facility is in the tropics with high ambient humidity throughout the year. Alongside detailed information on topography and plant layout, LNG vapour concentration reading from gas detectors were available for one time period from a recent unplanned cold venting event. The atmospheric conditions at the facility during the incident were available from nearby weather stations and used in the modelling: (i) air temperature during the incident was about 33 °C, (ii) relative humidity of 90%, (iii) wind speed was constant at 2 m/s at 10 m height and (iv) wind was blowing from the sea carrying the release gases towards the area of interest.

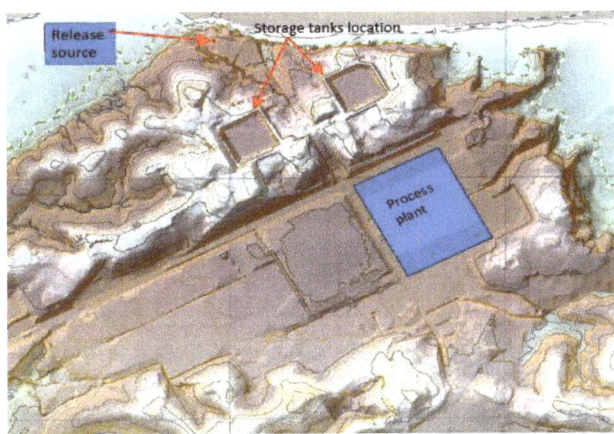

Figure 1. Schematic diagram showing the layout of key items. North is approximately top of the figure.

2.2. Computational Methods

As mentioned above, integral models were not adequate for this study. This is discussed in detail later. Integral models were used in the initial assessment of this incident and were found to be inadequate in explaining the event.

The computational fluid dynamics (CFD) code STARCCM+ was used for this study. The geometric model was constructed using the CAD data of the plant. Outside the confines of the CAD model, other generic data were used to include the vent stack, terrain, the two storage tanks explicitly. The dimensions and locations of these additional blockages were matched closely to the actual facility and used to calculate the drag source term and capture accurately their impact on the LNG dispersion.

2.2.1. Domain

The CFD calculation domain encompassed an area of 5 km^2. This area took into account sufficient distance upstream, for realistic flow establishment at the vent stack, and downstream to beyond the concentration of interest.

2.2.2. Gridding

Owing to the large area being considered, the number of grid cells was minimized using multi-block grid refinement techniques. Local grid refinements were used to ensure that the plume shape and behaviour was accurately captured (see Figure 2). In total, 8 million grid cells were needed, comprising of a mixture of prisms, hexahedra and polyhedral.

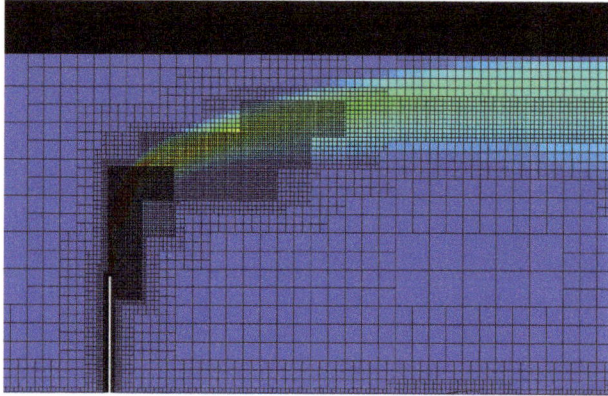

Figure 2. An example of grid layout across and close to the vent stack. The colour of contours of concentration is not important here and is shown for illustration.

2.2.3. Subgrid

For the purpose of this study, resolving the entire LNG plant was not necessary for the CFD dispersion analysis. Although the plant is over half a kilometre away, the blockages created by plant equipment and supporting structures would have an impact on the plume behaviour and on the ground-level concentrations near the area of interest. To resolve these items explicitly would render a vast increase in grid cell number and computer runtime. The two large tanks and large pieces of equipment were explicitly resolved. The effect of all process equipment, structures and pipework in the process area were represented in a 'subgrid scale' manner, i.e., their 'blockage' effects on the flow were modelled through local drag terms in the momentum equations derived from the properties of items from the CAD model for each grid cells. The drag terms were directional, e.g., for a subgrid scale pipe, there was negligible drag force along its length. Figure 3 presents the locations where subgrid scale modelling was used.

Figure 3. CFD geometric model showing smaller elements modelled on a subgrid scale basis with source terms applied locally.

2.2.4. Boundary Conditions

The boundary conditions used in the CFD modelling included vertical inlet wind profiles for a neutral atmospheric stability class, the appropriate velocity and turbulence factors (kinetic energy and dissipation rate). For the outlet of the computational domain, flow split outlet conditions was assumed.

2.3. Effect of Water Moisture

The cold LNG vapour was at $-161\ °C$, well below the freezing point of water. Moisture in the air would freeze forming a particulate suspension, increasing the effective density of the plume. This affected its behaviour and thus the concentration levels near the process areas, which were of interest. As well as dispersion, freezing, condensation, melting and evaporation of ambient air moisture were also modelled. It was also assumed that the various phases of water be treated as gaseous components with the appropriate composite molecular weight. This was to ensure that all the phases had the same velocity and temperature in each computational cell and that the density effects of the liquid and solid water were accounted for and affected the plume behaviour. Mass fraction of liquid water and ice that was present in each cell was small enough to be treated as dense gases. The mass and energy transfers between the phases in the entire CFD model based on local flow characteristics. We encountered numerical stability problems on these transfers and instigated numerical measures to control it. Figure 4 presents the plume iso-surfaces of ice and liquid water for a mass fraction of 0.001. For this study, relative humidity was assumed to be 90%, corresponding to a facility located by the sea in the tropics.

2.4. Topographical Effect

The bottom boundary of the calculation domain followed the contour of the terrain at the facility. The following conditions were applied: (i) no slip, (ii) rough wall boundary condition with an appropriate roughness value to account for the terrain vegetation distribution and (iii) adiabatic thermal boundary.

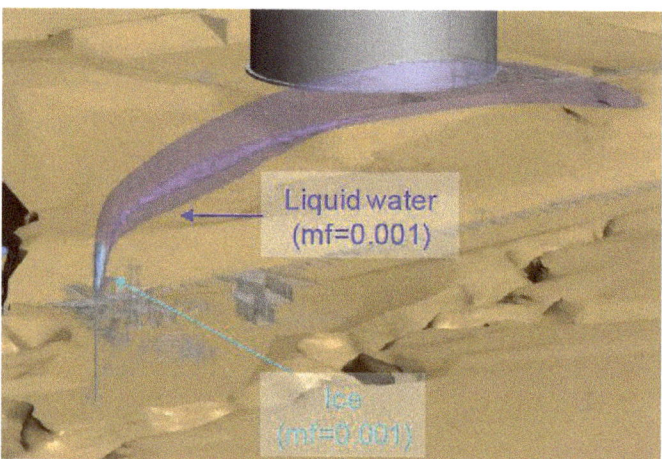

Figure 4. Plume iso-surfaces to represent ice and liquid water for a mass fraction (mf) of 0.001.

2.5. Steady State

As with integral models, this study focused on a steady-state solution. The total calculated flammable vented gas volume was used as an indicator for steady state (or computational convergence). Results were taken only when this has stabilised for a continuous and sufficient number of timesteps (>1000).

2.6. Venting Rates

Four venting rates were used in the study to represent various stages of the venting incident: 16 kg/s, 25 kg/s, 30 kg/s and 45 kg/s. The LNG vapour was assumed to have the physical properties of methane at a molecular weight of 17.2 and at a temperature of −161 °C, discharged at the top of the 20-inch diameter stack at height of 35 m above local ground level.

2.7. Wind Directions

The normal practice would be to align the wind direction towards the area of interest, in this case, the nearest point in the process plant. It was the hazardous and occupied area with potential for ignition. Four other wind directions about this direct alignment direction were also investigated and these are shown in Figure 5. The 5 wind directions assessed were representing wind blowing from 316° N to 338° N. These directions were chosen to be at or near the point where the plume could be obstructed or deflected by the 2 large storage tanks situated south of the flare stack but north of the process area. These storage tanks have measurable impact on the dispersion and impact on ground level gas concentration in the process area downwind.

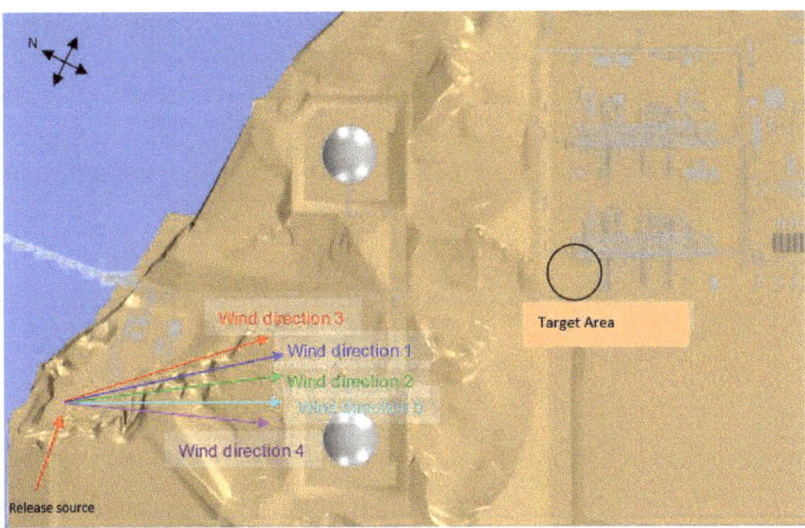

Figure 5. Schematic diagram showing the four wind directions. A target area on the first LNG train is also shown; the gas concentration calculated for this area will be used for comparison.

3. Results

Figure 6 represents a general 3D plot of the plume dispersing in the direction towards the process train. The colour of the iso-surfaces represents different levels of LFL concentration in the plume. As expected, the highest gas concentration would be immediate at the vent outlet and the lowest gas concentration would be at the farthest distance from the vent. The plume rose, levelled off and after a distance, descended onto the ground along which the plume continued to disperse. A summary of results is given in Table 1, which shows the effect of release rates, wind directions and wind speeds.

Figure 6. A 3D depiction of the calculated plume envelops for 4 gas concentrations (100% LFL, 50% LFL, 20% LFL and 10% LFL).

Table 1. Summary table of results showing the ground level concentration at the target area shown in Figure 5.

Case No.	Vent Rates (kg/s)	Wind Directions	Wind Speed (m/s)	Concentration (% LFL)
1	25	1	2	3.7
1a	30	1	2	4
1b	45	1	2	12.5
2	25	2	2	5.5
2a	30	2	2	6.3
3	25	3	2	3.8
3a	30	3	2	3
4	16	4	3.9	0
5	16	5	3	3.7
6	16	5	3.9	0

In general, the results showed that the vent stack was sufficient to disperse the LNG vapour sufficiently that it did not pose a flammable hazard on the plant, even at a high vent rate of 45 kg/s. The ground level concentration would have been zero using integral jet dispersion model. However, the focus of this study was less focused on the design adequateness but rather to explore factors that could have significant impact on the dispersion behaviour.

There were results that showed trends that were consistent with current common understanding of dispersion; the increase in target concentration occurred as release rate also increased (Cases 1, 1a and 1b).

There were also contrary results too. Concentration at the target area increased as wind direction deviated away from that which directly aligned with the vent and the target area. This is an example of effects of interaction between the wind field and terrain and the two large storage tanks.

3.1. Variation of Vent Rates

This section summarises the results of Cases 1, 1a and 1b where the vent rates ranged from 25 to 45 kg/s at a constant wind speed of 2 m/s from direction 1. Figure 7 represents the plume extent from Case 1, venting rate of 25 kg/s with 2 m/s wind from direction 1. The plume could be seen to meander downwards and in between both large storage tanks. The mean concentrations on the ground level were also captured. In order to compare gas measurements obtained during the incident with the CFD prediction, mean concentrations were also obtained from the modelling in terms of five concentric 10 m bands close to the position where the field measurements were obtained. The predicted values were about 3.7% LFL as compared to the minimum of 2% LFL measured on site (see Figure 8).

Case 1a is for the high release rate case of 30 kg/s, the effect of the topography could be seen with the plume splitting up into two, close to the vent exit and then descending onto the ground at two locations. This is shown clearly in Figure 9.

Again, mean concentrations near gas measurements were represented in five concentric 10 m bands. The predicted values were around 4% LFL as compared to the minimum of 2% LFL measured on site.

Figure 10 show the plume extent for Case 1b, the largest flowrate modelled of 45 kg/s and the ground concentration distribution at the process area. At this larger flowrate, it could be observed that a larger gas cloud reached the process area and touched down at two separate locations. Again, mean concentrations were also monitored at five concentric 10 m bands close to the position where gas measurements were obtained, and the predicted values were around 12.5% LFL compared to the minimum of 2% LFL measured on site.

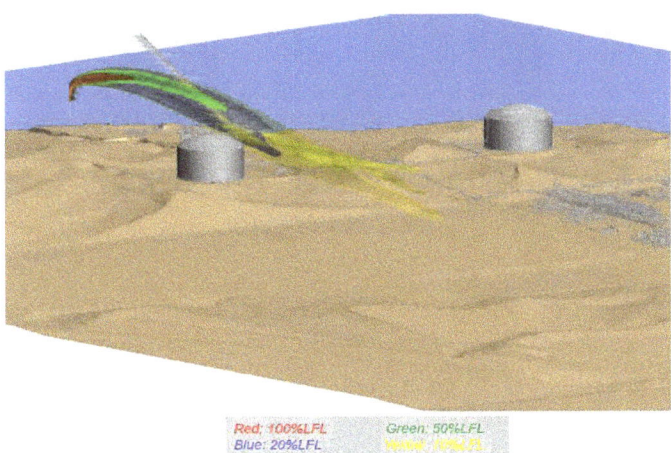

Figure 7. Case 1: iso-surface of 25 kg/s plume, 2 m/s wind, direction 1.

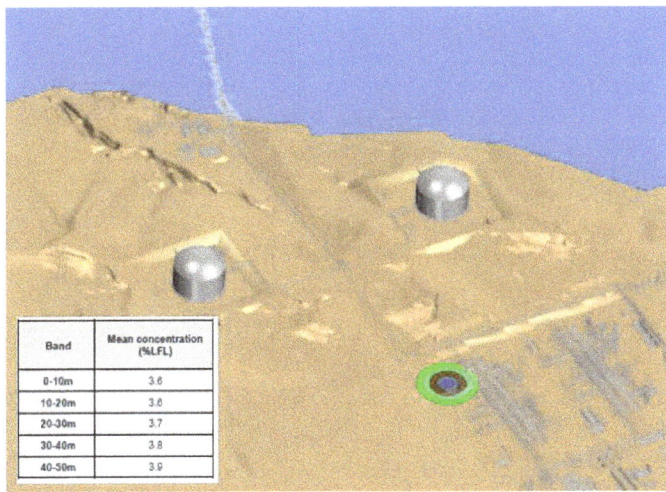

Figure 8. Case 1: concentric 10 m bands showing concentrations at gas detector location. The blue circle at the centre corresponds to the first 10 m band, the grey circle 10 to 20 m band, etc. The concentrations are the mean values in each band.

In general, ground level concentrations tended to increase in the process area with increasing vent rate. Low vent rate could lead to earlier touchdown of plume as observed in the iso-surfaces above. Venting rate of 25 kg/s produced a higher ground level concentration closer to the storage tanks than higher vent rates. This is summarised in Figure 11, which shows the ground level concentration distribution and touchdown locations for all 3 vent rates at 2 m/s from direction 1. The maximum ground level concentration at any location on the facility was about 10% LFL, which was below the facility point detectors low trigger set point of 20% LFL. The concentration at the location of interest and LOS detectors were estimated to be about 4% LFL and whilst this concentration was lower, it could trigger the LOS detectors. Note that during the incident, an equivalent reading of 2 to 15% LFL was captured by the LOS detectors.

Figure 9. Case 1a: iso-surface of 30 kg/s plume, splitting and touching down at two locations leading to different ground concentrations.

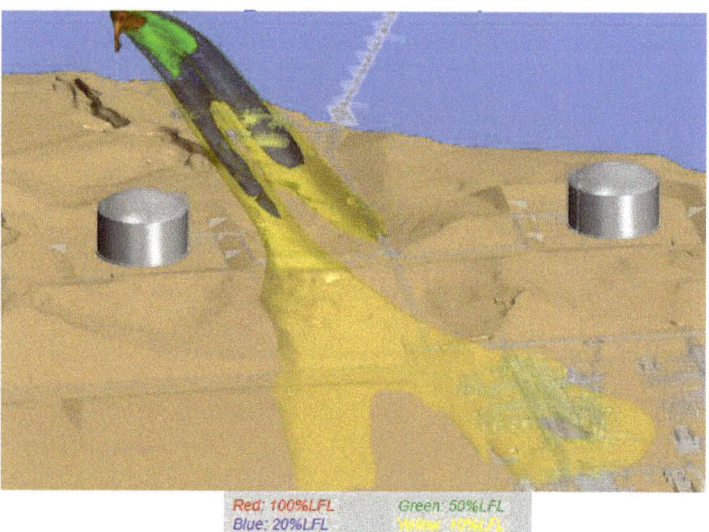

Figure 10. Case 1b: iso-surface of 45 kg/s plume, 2 m/s wind, direction 1.

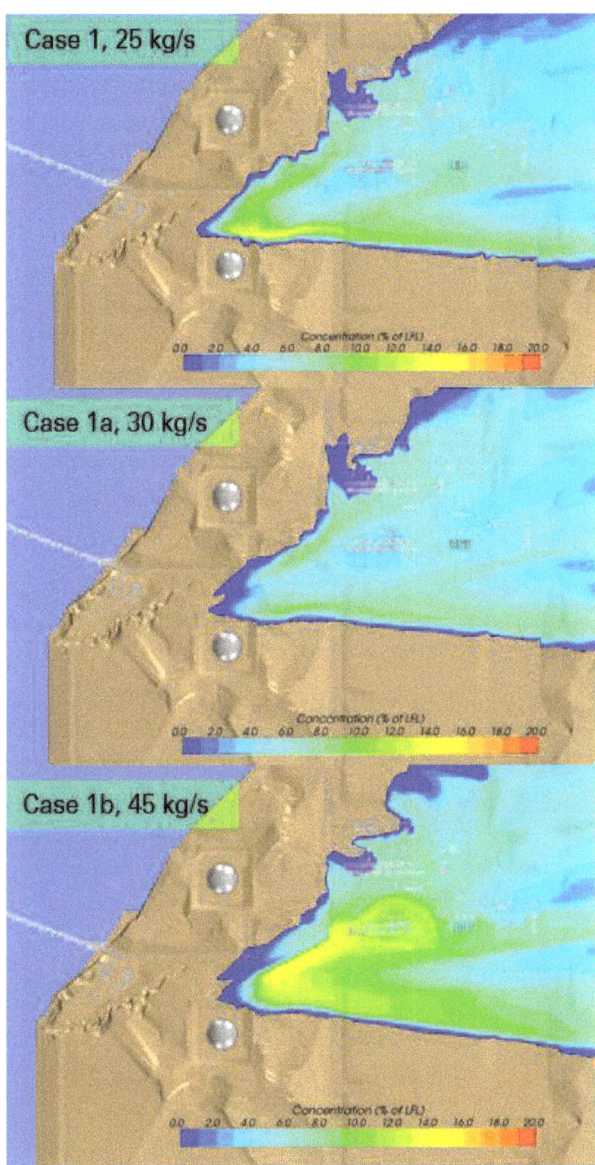

Figure 11. Corresponding ground level LFL concentration for a 25, 30 and 45 kg/s releases at a wind speed of 2 m/s and wind direction 1, showing the touchdown locations and ground level plumes.

3.2. Variation of Wind Directions

Wind direction 2 was modelled to be slightly farther clockwise of wind direction 1, going towards the west of the facility. Figure 12 shows the plume extent for Case 2, 25 kg/s release under 2 m/s wind. The plume was observed to potentially encroach on the process areas. This was in contrast with Case 1 (see Figure 7) where the plume dispersed more towards the east side of the facility. The mean concentrations monitored in 5 concentric 10 m bands close to the location of the LOS detectors were predicted to be about 5.5% LFL as compared to the minimum of 2% LFL measured on site.

Figure 12. Case 2: iso-surface of 25 kg/s plume, 2 m/s wind, direction 2.

Vent rate of 30 kg/s was also modelled under the same wind conditions. Figure 13 shows the extent of the plume for Case 2a where the plume concentration of 10% LFL covered a smaller area of the process facility as compared to the plume of 25 kg/s. Due to the higher flowrate, the same wind speed was less effective in dispersing the gas cloud, i.e., the higher concentration band of 20% LFL for Case 2a can be seen to be larger than Case 2. The mean concentration on the ground in the concentric bands were observed to be about 6.3% LFL.

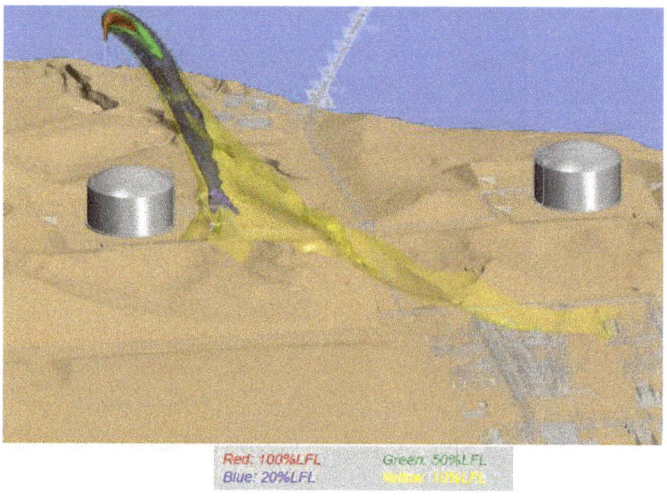

Figure 13. Case 2a: iso = surface of 30 kg/s plume, 2 m/s wind, direction 2.

Wind direction 3, anticlockwise of direction 1, going towards the east of the facility, was also modelled for the same flowrates of 25 and 30 kg/s. The plumes extended more towards to the east of the facility therefore led to lower mean ground concentration at the area of interest. The concentric bands measured 3.3 and 3% LFL for 25 and 30 kg/s releases, respectively. This was lower in comparison to the concentrations of 5.5 and 6.3% LFL for direction 2. See Figure 14 for a summary of ground level concentration for wind directions 2 and 3.

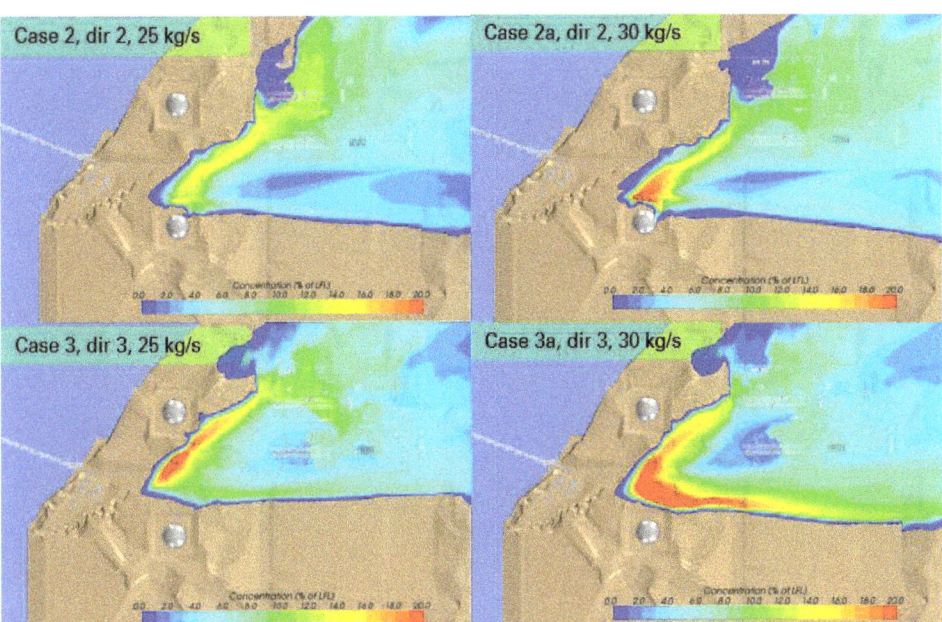

Figure 14. Corresponding ground level LFL concentration for a 25 and 30 kg/s releases at 2 m/s, direction 2 and 3, showing the touchdown locations and ground level plumes.

Wind directions 4 and 5 were also modelled for the same venting flowrate of 16 kg/s. The wind speeds were varied between 3.9 m/s and 3 m/s (see Figure 15). The ground concentration plots showed that for the higher wind speeds of 3.9 m/s, the plume did not reach the area of interest for wind direction 4 (Case 4) and reached the end section of the first LNG train for wind direction 5 (Case 6). However, as the wind speed decreased (Case 5), the plume touched the ground earlier and mean concentrations of about 3.7% LFL were observed near the LNG trains.

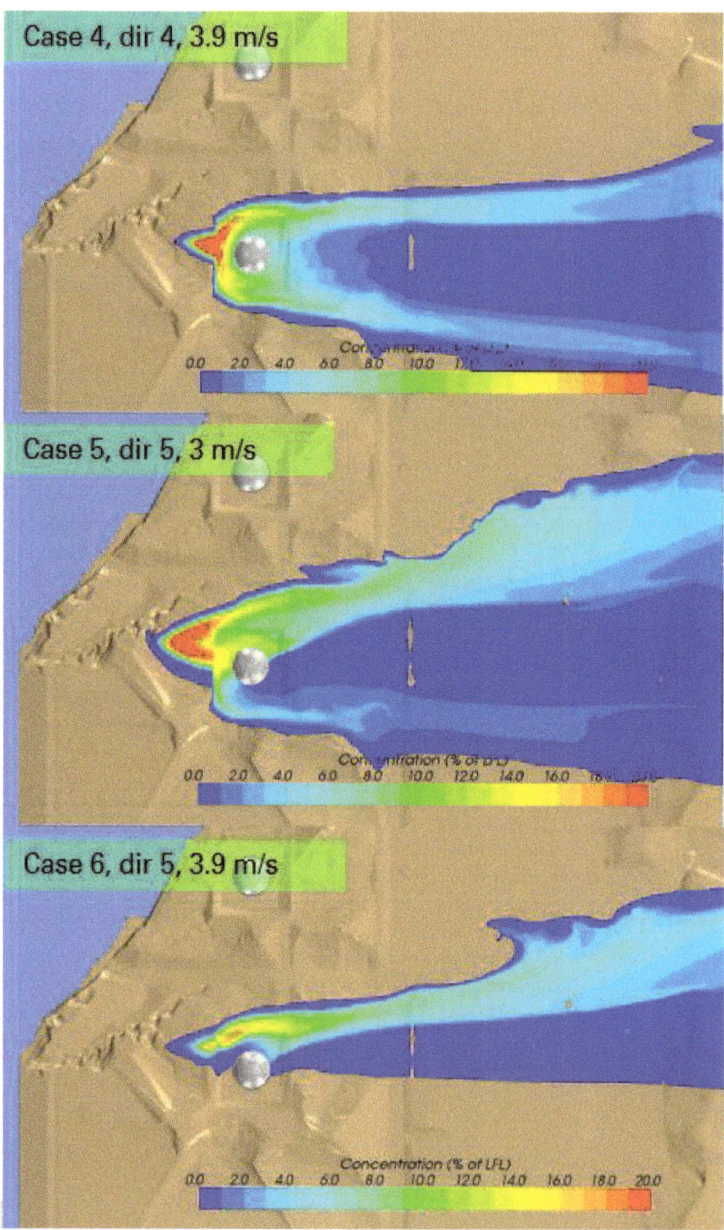

Figure 15. Ground level LFL concentration for 16 kg/s vent rate at different wind directions and speeds.

4. Discussions

The results of the CFD analysis showed that the likelihood of the plume touching down did not increase with vent rate, which was counterintuitive. As the release rate from the vent decreased, the concentration at the target sensor increased. This was the condition that encouraged the gas plume to touch down early in low wind and low release rate conditions. At higher wind speed, the plume dispersed aloft sufficiently that it did not descend to ground. No ground level concentrations were detected.

4.1. Comparison with Measurements

Readings from the log of gas detectors indicated that the LNG plume had touched down. This aspect agreed with the CFD results here. The reading further indicated that the gas concentration was between 2% LFL and 15% LFL. The calculated figures fell within this range (Table 1).

4.2. CFD vs. Integral Model

Prior to carrying out this CFD study, analysis using integral dispersion models were used. This included a jet dispersion model for an elevated source and a dense gas dispersion model for a low momentum source. This was conducted immediately after the incident with time constraint and limited field data for comparison. However, as mentioned in the section above on model assumptions, the initial calculations using integral jet dispersion model showed that the LNG plume continued to rise after its release and remained aloft throughout. The calculated ground level concentration consequently was very low. When the heavy LNG vapour was assumed to descend to the ground and then dispersed, the calculated concentration level at the distance of the target area was about 30% to 40% LFL for a vent rate of between 20 and 40 kg/s (compared to 2% to 15% LFL gas detector readings).

These calculations showed that commonly used integral jet dispersion model underestimated the flammable hazards as it did not predict the descend of the plume. The dense gas model, by not accounting for the initial momentum mixing, over-estimated the flammable hazards. It was recognised that the integral model was not applicable in this situation because: (i) the topography was not flat as there was a hill between the vent and storage tanks; (ii) there were two large storage tank in the path of dispersed vapour, before the process area, thus the topography and equipment could drag the plume towards the ground; and (iii) the very cold temperature of the vented gas could have condensed moisture in the atmosphere increasing the effective density of the plume. These effects will be described further below. It is therefore, in situations like those described in this paper, advisable to use CFD analysis.

4.3. Moisture Effect

Moisture in the air could affect the plume dispersion at various stages of the plume. The condensation and freezing led to formation of ice particles and release of latent heat. As the plume entrained warmer air, ice melted and then evaporated, absorbing heat and cooling the plume in the process. The mean plume temperature along its trajectory deviated from that when there was no moisture. The evolution of different phases of moisture in the plume is shown in Figure 4.

It is common practice in dispersion calculations to ignore the effect of moisture in air [4,5]. An alternative simpler approach had been used. Rather than modelling the evolution of moisture during dispersion, a modifier of ambient air properties could be used instead [6].

In a recent CFD study on Burro and Coyote LNG spill tests carried out in deserts in the USA, the effects of moisture in air on dispersion behaviour was found to be significant [7]. At 5% relative humidity (RH), the difference in calculated concentration at a location for including or excluding moisture effect was relatively small (~10%). However, the difference quickly rose to about 30% at an RH of 22%.

4.4. Good Match with Visible Plume (Not a Reliable Measure of Flammable Plume Shape)

It was tempting to use the visible plume to inform oneself of the flammable hazard distances or predict position of the plume when it descended to the ground, or whether it would descend to the ground. As can be seen in Figure 16, the calculated plume shape matched well with the observed visible plume; however, it was difficult to deduce the complete plume trajectory based on visible plume information.

Figure 16. Calculated plume compared with a picture of visible plume providing qualitative comparison.

Furthermore, the visible plume was not a good indicator of flammable extent as it would be dependent on RH [8]. For the plume section that was aloft, it was the RH local to the plume which may vary with height and locations; it was highly unlikely to be the same as that measured on the ground.

4.5. Ground Effect

Topographic effect was very evident from the results. It altered the wind velocity field and its distribution in the entire calculation domain. This resulted in the splitting up of the plume leading to two touch-down points and meandering ground level plumes, the trajectories of which were determined by the ground contours and the wind field (see Figure 6). This was consistent with the long standing guidance for environmental assessment [9].

4.6. Effect of Storage Tanks

Large objects, such as storage tanks, had similar effect as topography, but the effects were local. These large objects could induce downdraft, dragging the plume downward, promoting earlier touchdown or increased ground level concentration. This effect could be seen in Case 4 and 5 of Figure 15 where higher concentrations of vented gas was observed at the upstream side of the large storage tanks.

4.7. Plume Touchdown Location

The location of plume touchdown would affect the location and size of the hazard zone. A higher ambient wind increased the distance of the touchdown zone, giving a higher gas concentration at ground level farther downstream of the facility, rather than the target of interest or locations closer to the vent stack (see Table 1). For example, at 3 m/s wind speed in direction 5, the ground level concentration at the target area is 3.7% LFL, this is reduced to 0 at a higher wind speed of 3.9 m/s.

As the above results show, touchdown locations depend on the complex interactions between wind speed, wind direction with topography and equipment.

4.8. Further Work

As environmental conditions and vent rates changed with time, the next step is to assess these effects. It was deemed possible that during the analysis that the range of wind speeds and directions could be larger than those suggested by steady-state conditions. This is the subject of a separate paper that is in preparation.

4.9. Other Scenarios

This study addressed scenario which was not routinely assessed. There are other scenarios where LNG vapour can be generated at height, and they are beginning to be studied. This included the EU funded project SafeLNG that considers the vapour generation and its dispersion following a release of LNG at height which might have occurred after the rupture of an LNG import or export pipework at the top of an LNG tank [10].

5. Conclusions

This paper describes a CFD study of cold venting of LNG vapour. The integral dispersion model was found to be inadequate as many of its assumptions are not met, e.g., perfectly flat terrain.

The cold vapour induced phase change of ambient moisture, leading to changes in the thermodynamics as the vapour dispersed. This affected the dispersion and trajectory of the plume, i.e., aloft time, distance, touch down location and local ground concentration.

Topography and equipment altered the wind velocity field in the entire calculation domain, e.g., storage tanks downstream of the vent produced downwash effects, dragging the plume aloft down towards the ground, promoting earlier touchdown or increasing ground level concentration.

The analysis also showed the effects of different release rates, wind directions and wind speeds. There were results which showed trends which were as expected, i.e., increase in target concentration as release rate increased, but also contrary to expectation, i.e., concentrations at target area increased as wind direction deviated away from that directly aligned with the vent and target areas.

The results from this analysis showed broadly good agreement with key observations.

Author Contributions: Conceptualization, V.H.Y.T.; methodology, F.T., V.H.Y.T. and C.S.; formal analysis, F.T., C.S.; investigation, F.T., V.H.Y.T., C.S.; resources, C.S., F.T.; data curation, F.T.; writing—original draft preparation, V.H.Y.T.; writing—review and editing, F.T., C.S.; supervision, V.H.Y.T.; project administration, V.H.Y.T. (Key: F.T.—Felicia Tan, V.H.Y.T.—Vincent H. Y. Tam, and C.S.—Chris Savvides). All authors have read and agreed to the published version of the manuscript.

Funding: This research received no external funding.

Institutional Review Board Statement: Not applicable.

Informed Consent Statement: Not applicable.

Data Availability Statement: Not applicable.

Acknowledgments: Authors would like to thank Christophe Mabilat (then Atkins UK) for the generation of results and figures in this paper.

Conflicts of Interest: The authors declare no conflict of interest.

Abbreviations

CFD	Computational Fluid Dynamics
EU	European Union
LFL	Lower flammability limit
LOF	Line of Sight
LNG	Liquefied natural gas
RH	Relative humidity

References

1. Spicer, T.O.; Havens, J.A. Field test validation of the degadis model. *J. Hazard. Mater.* **1987**, *16*, 231–245. [CrossRef]
2. Ooms, G. A new method for the calculation of the plume path of gases emitted by a stack. *Atmos. Environ.* **1973**. [CrossRef]
3. Oil and Gas UK. *Fire and Explosion Guidance*; Oil and Gas UK: London, UK, 2007.
4. Hansen, O.R.; Gavelli, F.; Ichard, M.; Davis, S.G. Validation of FLACS against experimental data sets from the model evaluation database for LNG vapor dispersion. *J. Loss Prev. Process Ind.* **2010**, *23*, 857–877. [CrossRef]

5. Fardisi, S.; Karim, G.A. Analysis of the dispersion of a fixed mass of LNG boil off vapour from open to the atmosphere vertical containers. *Fuel* **2011**, *90*, 54–63. [CrossRef]
6. Cormier, B.R.; Qi, R.; Yun, G.; Zhang, Y.; Sam Mannan, M. Application of computational fluid dynamics for LNG vapor dispersion modeling: A study of key parameters. *J. Loss Prev. Process Ind.* **2009**, *22*, 332–352. [CrossRef]
7. Zhang, X.; Li, J.; Zhu, J.; Qiu, L. Computational fluid dynamics study on liquefied natural gas dispersion with phase change of water. *Int. J. Heat Mass Transf.* **2015**, *91*, 347–354. [CrossRef]
8. Vílchez, J.A.; Villafañe, D.; Casal, J. A dispersion safety factor for LNG vapor clouds. *J. Hazard. Mater.* **2013**, *246*, 181–188. [CrossRef] [PubMed]
9. Steven, R.; Hanna Gary, A.; Briggs Rayford, P.; Hosker, J. Handbook on atmospheric diffusion; prepared for the U.S Department of Energy. *Atmos. Environ.* **1983**. [CrossRef]
10. Wen, J.X. *Advances in Consequence Modelling for LNG Safety: Outcomes of the SafeLNG Project*; FABIG: Silwood Park, UK, 2018; pp. 25–32.

Article

Accounting Greenhouse Gas Emissions from Municipal Solid Waste Treatment by Composting: A Case of Study Bolivia

Magaly Beltran-Siñani and Antonio Gil *

INAMAT2—Science Department, Los Acebos Building, Campus of Arrosadia, Public University of Navarra, 31006 Pamplona, Spain; magaly.beltran@biosolarenergy.org
* Correspondence: andoni@unavarra.es; Tel.: +34-948-169-602

Citation: Beltran-Siñani, M.; Gil, A. Accounting Greenhouse Gas Emissions from Municipal Solid Waste Treatment by Composting: A Case of Study Bolivia. *Eng* **2021**, *2*, 267–277. https://doi.org/10.3390/eng2030017

Academic Editor: George Z. Papageorgiou

Received: 28 May 2021
Accepted: 24 June 2021
Published: 30 June 2021

Publisher's Note: MDPI stays neutral with regard to jurisdictional claims in published maps and institutional affiliations.

Copyright: © 2021 by the authors. Licensee MDPI, Basel, Switzerland. This article is an open access article distributed under the terms and conditions of the Creative Commons Attribution (CC BY) license (https://creativecommons.org/licenses/by/4.0/).

Abstract: Waste generation is one of the multiple factors affecting the environment and human health that increases directly with growing population and social and economic development. Nowadays, municipal solid waste disposal sites and their management create climate challenges worldwide, with one of the main problems being high biowaste content that has direct repercussions on greenhouse gases (GHG) emissions. In Bolivia, as in the most developing countries, dumps are the main disposal sites for solid waste. These places usually are non-engineered and poorly implemented due to social, technical, institutional and financial limitations. Composting plants for treatment of biowaste appear as an alternative solution to the problem. Some Bolivian municipalities have implemented pilot projects with successful social results; however, access to the economic and financial resources for this alternative are limited. In order to encourage the composting practice in the other Bolivian municipalities it is necessary to account for the GHG emissions. The aim of the present study compiles and summarizes the Intergovernmental Panel on Climate Change (IPCC) guidelines methodology and some experimental procedures for accounting of the greenhouse gases emissions during the biowaste composting process as an alternative to its deposition in a dump or landfill. The GHG emissions estimation results by open windrow composting process determined in the present study show two scenarios: 38% of reduction when 50% of the biowaste collected in 2019 was composted; and 12% of reduction when 20% of the biowaste was composted.

Keywords: municipal solid waste; SWDS; composting process; DOC; IPCC guidelines; GHG emissions

1. Introduction

Climate change has become a crosscutting issue in the management and direction of public policies worldwide, and the waste sector is an important contributor reflected in the GHG inventories. In Bolivia, according to the Plurinational Authority of Mother Earth (APMT), methane (CH_4) generated at solid waste disposal sites is responsible for approximately 10% of the annual global anthropogenic greenhouse gas emissions [1]. Additionally, according to the National Statistics Institute of Bolivia (INE), the Municipal Solid Waste (MSW) generation was 1,600,938 tons in 2019, of which 88% is generated in the urban area with a generation rate of 0.53 kg per inhabitant-day [2].

The Solid Waste Disposal Sites (SWDS) in Bolivia are mainly dumps, being approximately 6.8% disposed in sanitary landfills, 4.1% in controlled dumps and 89.1% in dumps [3]; of which about 30% are close to bodies of water that are used for human consumption and irrigation; these unsustainable practices generate leachates (percolated liquids), pollution of water, soil and atmosphere, and GHG emissions that affect the population health of the country [4]. A possible way of mitigating the MSW problem is firstly the differentiated collection to later compost it and allocate its nutrients for agriculture and forestry, through domestic use in gardens or orchards, and municipal use in gardening, landscaping, and recovery of degraded areas [5].

According to the IPCC guidelines, the composting process in general is given in aerobic conditions where a large part of the degradable organic carbon (DOC) in the waste material is transformed to carbon dioxide (CO_2). CH_4 is generated in anaerobic sections of the composting process [6]. According to Ahn [7], the GHG monitoring results in well managed composting plants, which show that the CO_2 produced is biogenic, and CH_4 and N_2O gases production are negligible. However, if proper composting conditions are not managed, CH_4 and N_2O emissions could potentially increase.

2. Methodology

The methodological steps provide a resume of the IPCC guidelines for the waste sector and the experimental procedure to determine the composting process GHG emissions when a municipality has separate biowaste collection and composting process such as MSW management strategy. After showing the calculation methodology, a study case for the Bolivian context is presented.

2.1. GHG Emission Sources

The IPCC guidelines present an internationally-approved methodology for the national GHG emissions and removals calculation and reporting [8]. Up until today, there are two IPCC reference guidelines, 1996 and 2006, and there is a 2014 refinement to the 2006 report that does not include refinement to the biological treatment in the waste sector.

In some cases, the Global Warming Potential (GWP) in the 2006 guidelines (Fifth Assessment Report) increased in comparison with 1996, such as with CH_4, and this particularity makes the specification of the guide important to use when a country reports its GHG inventory. In addition, in order to reduce double emissions accounting and to improve the coherence and completeness of the inventory, the 2006 guideline reduces from six groups of GHG emission sources to four: energy, Industrial Processes and Product Use (IPPU), Agriculture, Forestry and Other Land Use (AFOLU), and the waste sector.

2.2. Estimation Method

According to the 1996 and 2006 IPCC guidelines, the estimation method for accounting the GHG emissions is given by the Equation (1), where AD is the activity data that considers human activity with coefficients, and EF are the emission factors that quantify the emissions or removals per unit activity; the EF varies from default values (Tier 1) until more estimation complex methods (Tier 3); the parties members of the United Nations Framework Convention on Climate Change (UNFCCC) will choose their tier depending of their national circumstances and data availability [9].

$$Emissions = AD \times EF \qquad (1)$$

The GHG emissions estimation from the waste sector compiles activity data on its generation, composition and management. Solid waste management takes into account its collection, recycling, disposal sites, biological and other treatments, and incineration and open burning options [10]; CO_2, CH_4 and N_2O emissions estimation are considered for the waste sector GHG accounting. According to the 2006 IPCC guideline, it considers SWDS, biological treatment of solid waste, incineration and open burning of waste, and wastewater treatment and discharge categories. The accounting for CO_2, CH_4 and N_2O emissions varies according to the source categories as detailed in the Table 1.

As this study has focus on composting GHG accounting; SWDS and biological treatment of solid waste categories are reviewed [10].

2.2.1. Solid Waste Disposal Sites

The Revised 1996 IPCC Guidelines describe the mass balance method (Tier 1) and the First Order Decay (FOD) method (Tier 2) for estimating CH_4 emissions from SWDS. In 2006, a guidelines Tier 1 is given by the FOD method because it produces more accurate estimates of annual emissions [12]. In the FOD method the DOC content decays slowly in a few decades where CH_4 and CO_2 are produced; usually waste in a SWDS produces high amounts of CH_4 for the first years after its deposition. In order to achieve acceptably accurate results, the collected or estimated data should consider waste historical disposals over a time period of 3 to 5 half-lives and use disposal of at least 50 years [13].

Table 1. GHG source and emissions categories related to solid waste included in the IPCC 2006 guidelines [11].

Category	GHG	Inclusion in the Emissions Report	Comments
1. Solid waste disposal	CO_2	No	Biogenic origin and net emissions are accounted for the AFOLU Sector.
	CH_4	Yes	Fugitive emissions derived from the anaerobic decomposition of waste.
	N_2O	No	Presumed insignificant.
2. Biological treatment of solid waste	CH_4	Yes	Considers CH_4 and N_2O emissions for composting process; and CH_4 emissions for biogas production. If the biogas generated is used to produce energy, it will be reported in the energy sector and its N_2O emissions are presumed negligible.
	N_2O	Yes/No	
3. Incineration and open burning of waste	CO_2	Yes	The GHG emissions from waste incineration with energy recovery are reported in the Energy Sector, in the other case are reported in the waste sector; Only CO_2 emissions from fossil origin must be reported.
	CH_4	Yes	
	N_2O	Yes	

For this category, Tier 1 estimation values are based on default activity data and parameters from the IPCC FOD method; Tier 2 use the IPCC FOD method and some default parameters; however, it requires good quality country-specific AD (statistics, surveys or other similar sources) for at least 10 years or more on historical waste in SWDS; Tier 3 is based on good quality country-specific AD and the FOD method is used with developed or measured country-specific parameters. In addition, another method with equal or higher quality to the Tier 3 method can be used [10].

Considering that CH_4 is generated with the organic material degradation under anaerobic conditions, in the cover of the SWDS part of the CH_4 is oxidized and can be recovered for energy or flaring. Given this fact, the CH_4 emissions from SWDS for a single year only consider the fraction of CH_4 that is not recovered and can be estimated with the Equation (2).

$$CH_4 \ Emissions = \left[\sum_x CH_4 \ generated_{x,T} - R_T\right] * (1 - OX) \quad (2)$$

where $CH_4 \ Emissions$ are the total CH_4 emissions in the year of reference T in generated CH_4; x is the waste category or type of material; R_T is the total amount of CH_4 recovered in the year of reference, OX is the oxidation factor (fraction) in the year of reference T [12]. CH_4 generation depends on MSW information (waste and SWDS types) and it can be determined for the following equations:

$$DOC = \sum_i (DOC_i * W_i) \quad (3)$$

$$DDOC_m = W * DOC * DOC_f * MCF \quad (4)$$

$$DDOC_{ma_T} = DDOC_{md_T} + \left(DDOC_{ma_{T-1}} * e^{-k}\right) \quad (5)$$

$$DDOC_{m \ decomp_T} = DDOC_{ma_{T-1}} * \left(1 - e^{-k}\right) \quad (6)$$

$$CH_4 \ generated_T = DDOC_{m \ decomp_T} * F * 16/12 \quad (7)$$

where DOC is the fraction of degradable organic carbon in bulk waste in Gg of C/Gg of waste given by Equation (3), that considers DOC_i as the fraction of DOC in waste type i; and W_i as the fraction of waste type i. The Decomposable Degradable Organic Carbon ($DDOC_m$) is defined by Equation (4), considering W as the waste mass deposited in Gg; DOC as the degradable organic carbon in the year of deposition Gg of C/Gg of waste in fraction; DOC_f as the DOC fraction that can be decomposed in fraction (recommended default value 0.5 considering that SWDS environment is anaerobic and DOC values include lignin); and MCF as the CH_4 correction factor in aerobic conditions in the year of deposition in fraction (default values provided by IPCC for SWDS managed in anaerobic conditions and unmanaged in less than 5 m of high are 1 and 0.4, respectively) [12,14].

Additionally, the FOD basis is the first order reaction, where the CH_4 generation only depends on the total mass of decomposing material currently in the site. For this reason, the FOD calculations can be done by the Equations (5) and (6), being $DDOC_{maT}$ and $DDOC_{maT-1}$ the $DDOC_m$ accumulated in the SWDS at the end of the year of reference T and $T-1$ respectively in Gg; $DDOC_{mdT}$ the $DDOC_m$ deposited in the SWDS in the year of reference T in Gg; $DDOC_{m\ decompT}$ the $DDOCm$ decomposed in the SWDS in the year of reference T in Gg; and k the reaction constant, for Tier 1, k values for tropical sites up than 20 °C are 0.07 for paper; 0.17 for garden and park waste; and 0.4 for food waste in most and wet conditions [13].

Finally, the CH_4 generation is given by the Equation (7), considering F as the fraction of CH_4 (vol/vol) generated in the landfill; and the factor 16/12 as the molecular weight ratio CH_4/C [14].

2.2.2. Biological Treatment of Solid Waste: Composting Process

Composting involves biological treatment where organic material is degraded through microorganisms; leading to the compost production that can be used as a natural fertilizer or to improve soil structure [15]. During the process oxygen availability, C/N ratio, humidity, and temperature are the most important parameters that should be controlled under its three phases: Thermophilic, maturing and cooling. Under the thermophilic phase the material is decomposed, and the pathogens and bacteria are reduced by high temperatures above 55 °C [16]; in the maturing phase the temperature decreases for the low biological activity; and in the cooling phase the material gets very stable and mature [17].

The microbial activities under anaerobic and aerobic conditions during the composting process leads to the production of CO_2, CH_4, N_2O, and NH_3, with the CO_2 and CH_4 production given by the insufficient diffusion of O_2 [18]; the N_2O production depends on the temperature, nitrate content and the aeration rate [19], and NH_3 production has a direct relation with the temperature and pH [20].

The IPCC methodology for biological treatment of solid waste is given in the 2006 guideline that includes CH_4 and N_2O emissions from compost preparation considering the Mechanical–Biological (MB) treatment and the composting process. MB treatment involves separation, shredding and crushing operations on the organic material; CH_4 and N_2O production during the MB treatment depend on the specific operation and the time process [14]. The CH_4 and N_2O emissions estimation can be determined by the following steps:

Step 1: Data collection on the amount of solid waste that is composted (regional and country-specific default data for some countries is given in the 2006 IPCC guideline) [14];

Step 2: Estimate the CH_4 and N_2O emissions from composting process with Equations (8) and (9). The EF must be considered according to the facilities to get the specific information (tiers).

$$CH_4\ Emissions = M * EF * 10^{-3} \quad (8)$$

$$N_2O\ Emissions = M * EF * 10^{-3} \quad (9)$$

where: $CH_4\ Emissions$ and $N_2O\ Emissions$ are the total CH_4 and N_2O emissions per year in Gg respectively, considering M as the mass of organic waste processed in Gg; and EF in g of CH_4/kg and g of N_2O/kg of waste treated, respectively.

The CO_2 emissions are not considered given its biogenic origin [6], CH_4 and N_2O GWP are 28 and 265 times higher than CO_2 respectively [21]. CH_4 generation can occur at the beginning of the composting process under anaerobic conditions [22]; and N_2O generation can take place at various stages of the process by-product of nitrification or de-nitrification.

The EF by default or Tier 1 for CH$_4$ and N$_2$O emissions, and its uncertainty are given in the Table 2, during mechanical operations the GHG emissions can be considered negligible. EF for Tier 2 should be based on applied country representative measurements during the composting process; and EF for Tier 3 would consider facility/site-specific measurements (on-line or periodic) that are more reliable than Tier 2.

Table 2. Default emission factors for CH$_4$ and N$_2$O emissions from composting process [14].

Type of Basis	CH$_4$ Emission Factors (g of CH$_4$/kg Waste Treated)	N$_2$O Emission Factors (g of N$_2$O/kg Waste Treated)	Remarks
On a dry weight	10 (0.08–20)	0.6 (0.2–1.6)	Assumptions: 25–50% DOC in dry matter. 2% N in dry matter, 60% moisture content.
On a wet weight	4 (0.03–8)	0.24 (0.06–0.6)	

2.3. Sampling Methods

Experimental composting measurements are oriented to collect more affordable information for the *EF* such as with Tier 2 and Tier 3. Two experimental methods for sampling GHG emissions from open windrow composting are reviewed: flux chamber and funnel.

2.3.1. Flux Chamber Method

The flux chamber method has been successfully used for measuring GHG emissions in composting piles with organic household wastes [23], which basically uses inverted boxes or cylinders (with known dimensions) situated over the compost pile surface where the gases produced concentration is measured by several instrumental techniques [24]. This method can be done by closed and open chamber types, and non-reactive materials (stainless steel, aluminum, PVC, polypropylene, polyethylene, or plexiglass) are recommendable to construct the flux chambers [25].

The closed chambers shapes usually are cylinders with 10 to 400 L of volume [23] where gases concentration are sampled from 10 to 30 min intervals depending on the instrumental technique. Gas analysis can be done on site (sample return into the chamber) in order to avoid pressure changes or off-site, being the samples stored and analyzed in the laboratory generally by gas chromatography technique. The *EF* determination is determined by the following equations:

$$E_{Flux\_chamber} = \frac{dC_{gas}}{dt} * V_{chamber} \qquad (10)$$

$$EF_{gas} = \frac{\sum \int_{t_2}^{t_1} E_{Flux\_chamber} * dt}{m_{input\_waste}} \qquad (11)$$

where, the gas emission fluxes ($E_{Flux\_chamber}$) in kg/h [18] considers dC_{gas}/dt as the change in the concentration C_{gas} over time in kg/s·m, and $V_{chamber}$ as the total volume inside the chamber in m^3. The emission factor of the gases (EF_{gas}) integrates over time (time between measurements) and summarize over the entire year of composting the $E_{Flux\_chamber}$ considering m_{input_waste} as the total input organic waste amount in Mg being its units kg/Mg.

This method is economic and easy doing as its main advantages. However, the pressure gradients between compost pore space and chamber headspace can be induced, the high height of the chamber may not allow adequate mixing of headspace air [25], and the rate of diffusion of gases can be perturbed or decreased leading to an underestimation of GHG emissions [24].

The open chambers allow the capture of the whole flux of gases generated by the compost process, where the measurements can be collected by different sections and depths of the pile including temperature and oxygen profiles. The daily gas flux from the top, upper or lower side of pile in g/d (*E*) can be calculated by Equation (12) considering

Q_{sweep} as the flow rate of the N_2 sweep gas going into the chamber in L/h; C_{sample} as the concentration of the gas in vol/vol determined by the gas chromatograph; Y_{sample} as the concentration of the gas in mg·L^{-1} that is converted from C_{sample} assuming ideal gas relations and using chamber air temperature values (measured by a thermocouple thermometer); A as the bottom of the chamber surface emissions area in m^2; and B as the top, upper or lower side of pile surface area in m^2 [26].

$$E = \frac{Q_{sweep} * \left(1/\left(1 - C_{sample}\right)\right) * Y_{sample}}{A} * B * 24 * 1000 \qquad (12)$$

The daily average mass-based GHG fluxes should be determined summarizing the E values of the entire pile divided by the biowaste input weight of the pile and the total composting days. The annual GHG emissions can be determined summarizing the cumulative gas emissions and the total weight of biowaste during a year of compost producing. The EF can be determined with the annual GHG emissions divided by the dry mass of biowaste.

The uncertainty can be determined by the standard deviation of the mean value from three replicates collected in each sampling event using Equation (13), considering A and B the standard deviation of the mean value from the three replicates sampled on day a and day $a + t$ respectively; k as the "coverage factor" with value $k = 2$ for confidence level of almost 95% [26].

$$Uncertainty = \sum \sqrt{\left(\frac{A}{2} * t\right)^2 + \left(\frac{B}{2} * t\right)^2} * k \qquad (13)$$

As the air flux rate is affected by the environmental conditions variation this method requires the flow control and correction for changes in temperature and atmospheric pressure being its main limitation [24].

2.3.2. Funnel Method

The success funnel method developed by the consulting group Ramboll is used for surface GHG emissions measuring in triangular compost windrows [17]; its measuring instrument can be made of aluminum and resembles where an upside-down funnel covers usually 1 m^2 of a windrow and a vent pipe is attached to the top of the funnel [27].

Convection is an important factor in open compost windrows, as the air flows through the windrow and transports gases away from it into the atmosphere [17], given this fact EF from open compost windrows are difficult to determine considering that usually a small surface is covered and it works as a static chamber that does not allow to measure the gas emissions via convection; in order to overcome this limitation and to improve the accuracy of gas emission estimation, Phong proposed a funnel method covering almost 50 m^2 of area, adding forced ventilation from one side of the funnel [17]. With this modification EF for each gas using the following equations:

$$f_{funnel} = \frac{C_{gas} * v_{air} * A_{vent\_pipe}}{A_{funnel}} \qquad (14)$$

$$E_r = \frac{E_{out} - E_{in}}{A_{funnel}} * Q_{funnel} \qquad (15)$$

$$EF_d = \frac{E_r * 24}{1000} * \frac{A_w}{M_w} \qquad (16)$$

$$EF = \sum EF_d * T \qquad (17)$$

The determination of EF in g/Mg considers T as the composting duration in days; E_{fd} as the emission factor per day in g/Mg·d; A_w as the total surface of the windrow in m^2; M_w as the total mass of the windrow in Mg; 24 and 100 as the correction factors from hours to days and from mg to g respectively; E_r as the emission rate in mg/h·m²; E_{in} and E_{out} as the concentrations of inlet and outlet in mg/m³; Q_{funnel} as the input air in m³/h; A_{funnel} as the surface under the tunnel in m²; C_{gas} as the gas concentration sampled from the chamber in vol/vol; v_{air} as the air flow velocity in the vent pipe in m/s; and A_{vent_pipe} as the sectional area of the vent pipe in m².

After GHG emissions calculation, the GHG reduction potential can be determined by the Equation (18), being E_{SWDS} and $E_{composting\_process}$ the total emissions from biowaste in SWDS and in composting plants respectively in Gg of $CO_{2\text{-eq}}$; and $E_{reduction}$ the total gas reduction for the composting process in Gg of $CO_{2\text{-eq}}$.

$$E_{reduction} = E_{SWDS} - E_{composting\_process} \qquad (18)$$

3. Results and Discussion

The results presented in this section consider the 2006 IPCC methodology as conducive to more affordable characteristics; in addition, as there is not available country specific EF (Tier 2 and Tier 3), Tier 1 EF are used for SWDS category.

In the Bolivian context, MSW information is available since 2003, being the input parameters given by official statistics organization from Bolivia and some default values available in 2006 IPCC guidelines, values that are presented in the Table 3. The $DDOC_{m\ decompT}$ for the period 2003 to 2019 is presented in the Figure 1, which shows the biowaste that is decomposed each year.

Table 3. Input parameters of solid waste disposal sites in Bolivia (self-elaboration).

No	Parameter	Symbol	Unit	Value	Remarks
1	Mass of the waste deposited in 2019	W	Gg	1 601	Official data of INE Bolivia [28]
2	Degradable organic carbon	DOC	fraction	0.11	Considering 55.2% of organic waste and 8% of paper [29]
3	Fraction of DOC that can decompose	DOC_f	fraction	0.50	IPCC default value [13]
4	CH_4 correction factor for aerobic decomposition	MCF	fraction	0.82	70% are disposed in controlled sites and 30% are disposed in unmanaged sites <5 m [4]
5	Reaction constant	k	fraction	0.23	Considering 55.2% of organic waste and 8% of paper; and default values of IPCC [13]
6	Fraction of CH_4 in generated landfill gas	F	fraction	0.5	IPCC default value [13]
7	Total amount of CH_4 recovered in 2016	R_T	Gg	0	There is no methane recovery in landfills [29]
8	Oxidation factor	OX	fraction	0	IPCC default value [13]

The GHG estimated emission from SWDS for the year 2019 is 40.67 Gg of CH_4, being that this value is in the range of the last Bolivian GHG inventory for the year 2008 (51.14 Gg of CH_4 calculation based on 1996 methodology) [30], the difference between these values can be related to the IPCC methodology use.

In Bolivia, pilot composting plants in some municipalities are processing biowaste since 2006 with 60% of efficiency, in other words, 0.4 tons of compost are produced per ton of biowaste [31]. The GHG emissions estimation from composting facilities is determined considering two scenarios: 50% and 20% of the biowaste collected during 2019 are composted respectively. As there is no experimental measured EF (Tier 2 or Tier 3) from the Bolivian composting plants, experimental EF from other sources are considered; additionally, IPCC default values (Tier 1) is take into account, these EF by method and the GHG emission from the composting process is given in the Table 4.

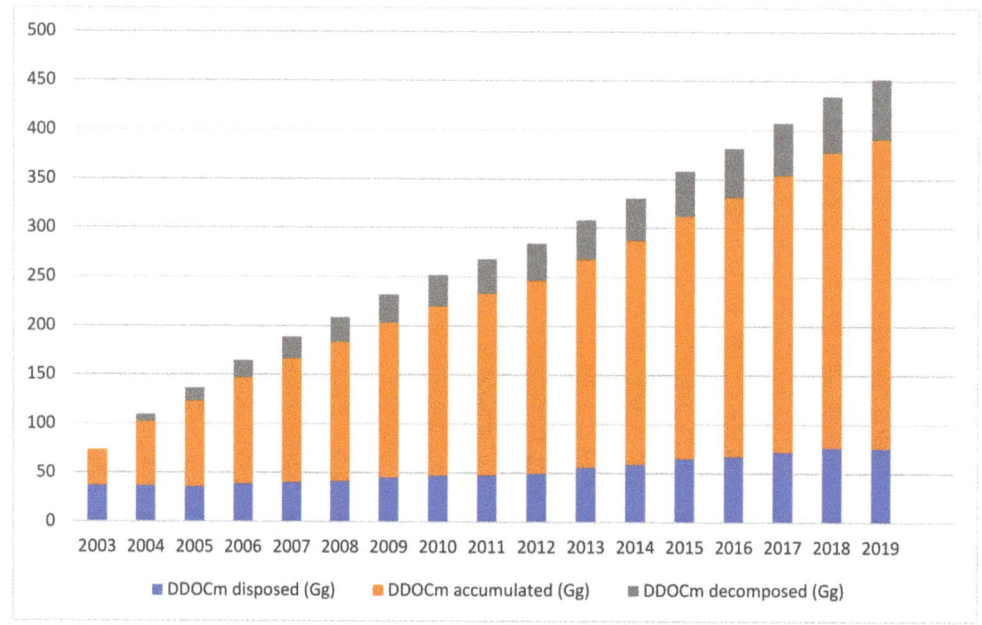

Figure 1. FOD method for solid waste disposal sites in Bolivia (self-elaboration).

Table 4. GHG emissions in open windrow composting plants of Bolivia (self-elaboration).

No	Biowaste Treated in Composting Plants	Weight Biowaste (Gg)	Method	CH_4 Emissions (Gg of CH_4)	N_2O Emissions (Gg of N_2O)	Total GHG Emissions (Gg of $CO_{2\text{-eq}}$)
1	50% of the organic waste collected during 2019	441.86	IPCC default values (Tier 1)	1.77	0.11	77.59
			Flux chamber: Closed chamber (Tier 2)	0.93	0.12	58.18
			Funnel (Tier 2)	0.90	0.01	28.34
2	20% of the organic waste collected during 2019	176.74	IPCC default values (Tier 1)	0.71	0.04	31.04
			Flux chamber: Closed chamber (Tier 2)	0.37	0.05	23.27
			Funnel (Tier 2)	0.36	0.00	11.33

The determined emissions in open windrow composting facilities shows that the IPCC default values are overestimated as was found by different authors [17]. Another aspect to consider is that the decomposition of biowaste in SWDS takes almost 100 years according to the IPCC guidelines, in comparison, the composting plants reduce the organic content in months with low GHG emissions, and its residual product is beneficial as a fertilizer for agriculture activities.

In order to show the GHG emissions reduction by the implementation of composting plants in Bolivia as a MSW treatment facility, the Figure 2 shows that over a lifetime of 100 years, the 1601 Gg of waste deposited in SWDS in 2019 generates 1067 Gg of CH_4 or 29,884 Gg of $CO_{2\text{-eq}}$ in total. However, if the first scenario is considered there is a reduction of 38% of GHG emissions in total, and for the second scenario there is a reduction of 12% (considering Tier 1 EF).

As the results of GHG emissions reduction by composting facilities show, it reduces significantly the GHG emissions and the volume of biowaste that would be cumulated in SWDS. As in Bolivia the main disposal sites are open dumps [3] the benefits to access to biological treatment of biowaste are valuable since the climate change point of view, in addition, its product properties as a fertilizer can be used for the agriculture activity that contributes around 15% of the national Gross Domestic Product (GDP) [32].

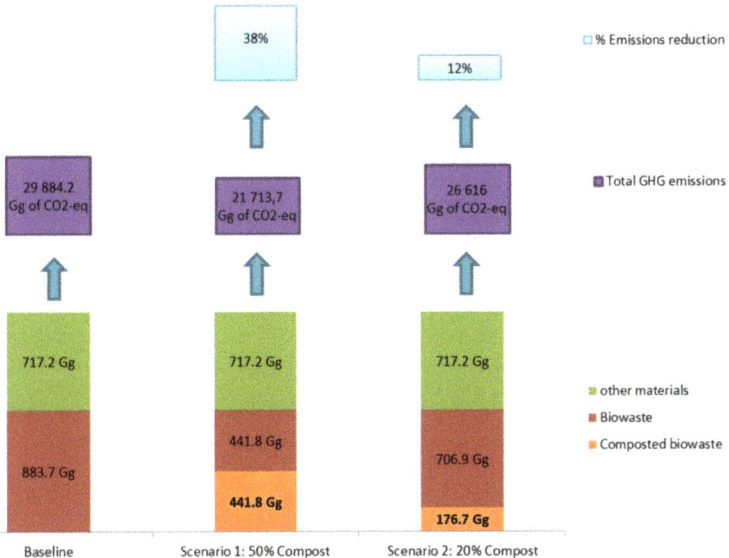

Figure 2. GHG emissions reduction by implementing composting facilities for MSW treatment (self-elaboration).

Finally, the main benefits of composting MSW are: the reduction of GHG emissions from the waste sector and the carbon footprint; as a fertilizer it can enrich the soil (compost helps to retain moisture and to suppress pests and plant diseases) promoting higher yields of agricultural crops (with the biodiversity increasing, reducing chemical fertilizer needs); the opportunity to get involved in humus (rich nutrient for plants) production; the potential reduction of leachate produced in SWDS, and the air quality improving (burning yard waste releases harmful chemicals into the air, producing diseases such as asthma); positive environmental and health impacts; and job opportunity creation as social and economic impacts.

Author Contributions: Conceptualization, M.B.-S. and A.G.; methodology, M.B.-S. and A.G.; formal analysis, M.B.-S.; investigation, M.B.-S.; writing—original draft preparation, M.B.-S.; writing—review and editing, M.B.-S. and A.G. All authors have read and agreed to the published version of the manuscript.

Funding: This research received no external funding.

Institutional Review Board Statement: Not applicable.

Informed Consent Statement: Not applicable.

Data Availability Statement: Not applicable.

Acknowledgments: M.B.-S. thanks Center for International Postgraduate Studies for Environmental Management for carry on the capacity building and the research program. A.G. also thanks Santander Bank for funding via the Research Intensification Program.

Conflicts of Interest: The authors declare no conflict of interest.

References

1. Ministerio de Medioambiente y Agua. Inventario de Gases de Efecto Invernadero de Bolivia 2002–2004. In *Programa Nacional de Cambios Climáticos en el Marco de la Segunda Comunicación Nacional de Bolivia Ante la Convención Marco de las Naciones Unidas Sobre el Cambio Climático (CMNUCC)*; Ministerio de Medioambiente y Agua: La Paz, Bolivia, 2017; Volume 1. (In Spanish)
2. Viceministerio de Agua Potable y Saneamiento Básico. Diagnóstico de la Gestión de Residuos Sólidos en Bolivia. Viceministerio de Agua Potable y Saneamiento Básico: La Paz, Bolivia, 2010. (In Spanish)
3. Morales, S. *MMAyA*; Ministerio de Medioambiente y Agua: La Paz, Bolivia, 2019.
4. Navarro, F.; Marco, B.; Massimo, Z.; Torretta, V.M.R. An Interdisciplinary Approach for Introducing Sustainable Integrated Solid Waste Management System in Developing Countries: The Case of La Paz (Bolivia). SNSIM. 2016. Available online: https://www.ecomondo.com/ecomondo/programma-eventi/procedia_esem_vol_3_nr_2_2016.pdf#page=20 (accessed on 20 January 2021).
5. Scow. Handbook Composting Market. 2016. Available online: http://www.biowaste-scow.eu/Handbook-for-Compost-marketing (accessed on 20 January 2021).
6. IPCC. Biological Treatment of Solid Waste. 2006. Available online: https://www.ipcc-nggip.iges.or.jp/public/2006gl/pdf/5_Volume5/V5_4_Ch4_Bio_Treat.pdf (accessed on 20 January 2021).
7. Ahn, H.K.; Mulbry, W.; White, J.W.; Kondrad, S.L. Pile mixing increases greenhouse gas emissions during composting of dairy manure. *Bioresour. Technol.* **2011**, *102*, 2904–2909. [CrossRef] [PubMed]
8. IPCC. How Does the IPCC Work? 2019. Available online: https://archive.ipcc.ch/organization/organization_structure.shtml (accessed on 20 January 2021).
9. UNFCCC. Guidelines for the Preparation of National Communications from Parties Not Included in Annex I to the Convention. 2011. Available online: https://unfccc.int/sites/default/files/17_cp.8.pdf (accessed on 20 January 2021).
10. IPCC. Guidelines for National Greenhouse Gas Inventories. 2015. Available online: https://www.ipcc-nggip.iges.or.jp/public/2006gl/vol5.html (accessed on 20 January 2021).
11. Cabrera, C.L. Emisiones de Metano Derivadas de los Desechos Sólidos Municipales en Cuba. *Obs. Ambient.* **2011**, *14*, 23. Available online: http://bbibliograficas.ucc.edu.co:2068/docview/963362173/6F0465626ABC4D55PQ/53?accountid=44394%0A (accessed on 20 January 2021). (In Spanish)
12. Riitta, P.; Wagner, J.; Silva, A.; Qingxian, G.; López Cabrera, C.; Katarina, M.; Hans, O.; Elizabeth, S.; Chhemendra, S.; Alison, S.; et al. Biological treatment of solid waste. In *IPCC Guidelines for National Greenhouse Gas Inventories*; IPCC: Geneva, Switzerland, 2006.
13. Riitta, P.; Svardal, P.; Wagner, J.; Silva, A.; Qingxian, G.; Carlos, L.C.; Mareckova, K.; Hans, O.; Masato, Y. Solid waste disposal. In *IPCC Guidelines for National Greenhouse Gas Inventories*; IPCC: Geneva, Switzerland, 2006; pp. 1–40.
14. Riitta, P.; Chhemendra, S.; Masato, Y.; Wagner, J.; Silva, A.; Qingxian, G.; Matthias, K.; López Cabrera, C.; Katarina, M.; Sonia Maria, M.V. Waste Generation, Composition. *IPCC Guidelines for National Greenhouse Gas Inventories*. 2006. Available online: https://www.ipcc-nggip.iges.or.jp/public/2006gl/ (accessed on 20 January 2021).
15. Plana, R. *Compostaje de Residuos Orgánicos*; Gobierno de Catalunya: Catalunya, Spain, 2015.
16. Lim, S.L.; Lee, L.H.; Wu, T.Y. Sustainability of using composting and vermicomposting technologies for organic solid waste biotransformation: Recent overview, greenhouse gases emissions and economic analysis. *J. Clean. Prod.* **2016**, *111*, 262–278. [CrossRef]
17. Phong, N.T. Greenhouse gas emissions from anaerobic digestion plants. *Vietnam J. Sci. Technol.* **2018**, *54*, 208–215. [CrossRef]
18. Hao, X.; Chang, C.; Larney, F.J.; Travis, G.R. Greenhouse gas emissions during cattle feedlot manure composting. *J. Environ. Qual.* **2001**, *31*, 376–386. [CrossRef] [PubMed]
19. Hellebrand, H.J. Emission of nitrous oxide and other trace gases during composting of grass and green waste. *J. Agric. Eng. Res.* **2008**, *69*, 365–375. [CrossRef]
20. Hansen, M.N.; Henriksen, K.; Sommer, S.G. Observations of production and emission of greenhouse gases and ammonia during storage of solids separated from pig slurry: Effects of covering. *Atmos. Environ.* **2006**, *40*, 4172–4181. [CrossRef]
21. Stocker, T.F.; Qin, G.-K.P.; Tignor, S.K.; Allen, J.B.; Nauels, Y.X.; Bex, V.; Midgley, P.M. *The Physical Science Basis. Contribution of Working Group I to the Fifth Assessment Report of the Intergovernmental Panel on Climate Change*; Cambridge University Press: New York, NY, USA, 2013.
22. Sánchez-Monedero, M.A.; Cayuela, M.L.; Roig, A. Strategies to transform organic residues from olive and wine industries: Greenhouse gas emissions and climate change. *Acta Horticul.* **2015**, *1076*, 57–68. [CrossRef]
23. Beck-Friis, B.; Pell, M.; Sonesson, U.; Jönsson, H.; Kirchmann, H. Formation and emission of N_2O and CH_4 from compost heaps of organic household waste. *Environ. Monit. Assess.* **2000**, *62*, 317–331. [CrossRef]
24. Sánchez, A.; Artola, A.; Font, X.; Gea, T.; Barrena, R.; Gabriel, D.; Mondini, C. Greenhouse gas from organic waste composting: Emissions and measurement. In CO_2 *Sequestration, Biofuels and Depollution*; Springer International Publishing: Cham, Switzerland, 2015; Volume 5, pp. 33–70.
25. Rochette, P. Towards a standard non-steady-state chamber methodology for measuring soil N_2O emissions. *Anim. Feed Sci. Technol.* **2011**, *166–167*, 141–146. [CrossRef]
26. Zhu-Barker, X.; Bailey, S.K.; Paw, U.K.T.; Burger, M.; Horwath, W.R. Greenhouse gas emissions from green waste composting windrow. *Waste Manag.* **2017**, *59*, 70–79. [CrossRef] [PubMed]
27. Andersen, J.K.; Boldrin, A.; Christensen, T.H.; Scheutz, C. Greenhouse gas emissions from home composting of organic household waste. *Waste Manag.* **2010**, *30*, 2475–2482. [CrossRef]

28. INE Bolivia. Solid Waste Generation. 2020. Available online: https://www.ine.gob.bo/index.php/medio-ambiente/residuos-solidos-cuadros-estadisticos/ (accessed on 20 January 2021). (In Spanish)
29. Ministerio de Medioambiente y Agua. *Guía Para el Aprovechamiento de Residuos Sólidos Orgánicos*; Ministerio de Medioambiente y Agua: La Paz, Bolivia, 2013.
30. Estado Plurinacional de Bolivia. Tercera Comunicación Nacional del Estado Plurinacional de Bolivia. 2020. Available online: https://unfccc.int/sites/default/files/resource/NC3%20Bolivia.pdf (accessed on 20 January 2021).
31. Quispe, M. Ministerio de Medio Ambiente y Agua: La Paz, Bolivia. 2019. Available online: https://www.mmaya.gob.bo/ (accessed on 20 January 2021). (In Spanish)
32. UDAPE. Sector Agropecuario Bolivia. 2004. Available online: https://www.udape.gob.bo/portales_html/diagnosticos/diagnostico2007/documentos/Documento%20Sector%20Agricola.pdf (accessed on 20 January 2021). (In Spanish)

Article

Adsorption of Estradiol by Natural Clays and *Daphnia magna* as Biological Filter in an Aqueous Mixture with Emerging Contaminants

Andrés Pérez-González [1], Verónica Pinos-Vélez [2,3,*], Isabel Cipriani-Avila [4], Mariana Capparelli [5], Eliza Jara-Negrete [4], Andrés Alvarado [3,6], Juan Fernando Cisneros [3,7] and Piercosimo Tripaldi [1,*]

1. Grupo de Investigación en Quimiometría y QSAR, Facultad de Ciencia y Tecnología, Universidad del Azuay, Cuenca 010204, Ecuador; aperez@uazuay.edu.ec
2. Departamento de Biociencias, Facultad de Ciencias Químicas, Universidad de Cuenca, Cuenca 010202, Ecuador
3. Departamento de Recursos Hídricos y Ciencias Ambientales, Universidad de Cuenca, Cuenca 010202, Ecuador; andres.alvarado@ucuenca.edu.ec (A.A.); juan.cisneros@ucuenca.edu.ec (J.F.C.)
4. Escuela de Ciencias Químicas, Pontificia Universidad Católica del Ecuador, Quito 170150, Ecuador; ecipriani111@puce.edu.ec (I.C.-A.); enjara@puce.edu.ec (E.J.-N.)
5. Facultad de Ciencias de la Tierra y Agua, Universidad Regional Amazónica Ikiam, Tena 150150, Ecuador; mariana.capparelli@ikiam.edu.ec
6. Facultad de Ingeniería, Universidad de Cuenca, Cuenca 010203, Ecuador
7. Departamento de Química Aplicada y Sistemas de Producción, Universidad de Cuenca, Cuenca 010203, Ecuador
* Correspondence: veronica.pinos@ucuenca.edu.ec (V.P.-V.); tripaldi@uazuay.edu.ec (P.T.)

Abstract: Among emerging pollutants, endocrine disruptors such as estradiol are of most concern. Conventional water treatment technologies are not capable of removing this compound from water. This study aims to assess a method that combines physicochemical and biological strategies to eliminate estradiol even when there are other compounds present in the water matrix. Na-montmorillonite, Ca-montmorillonite and zeolite were used to remove estradiol in a medium with sulfamethoxazole, triclosan, and nicotine using a Plackett–Burman experimental design; each treatment was followed by biological filtration with *Daphnia magna*. Results showed between 40 to 92% estradiol adsorption in clays; no other compounds present in the mixture were adsorbed. The most significant factors for estradiol adsorption were the presence of nicotine and triclosan which favored the adsorption, the use of Ca-montmorillonite, Zeolite, and time did not favor the adsorption of estradiol. After the physicochemical treatment, *Daphnia magna* was able to remove between 0–93% of the remaining estradiol. The combination of adsorption and biological filtration in optimal conditions allowed the removal of 98% of the initial estradiol concentration.

Keywords: wastewater treatment; natural clays; emerging contaminants; zeolite; bentonite; *Daphnia magna*; adsorption

1. Introduction

Currently, there is a growing interest in the so-called emerging contaminants (ECs), compounds of different origins and chemical nature whose presence in the environment is not considered significant in terms of distribution and/or concentration [1]. ECs include drugs and personal care products, surfactants, flame retardants, industrial and food additives, steroids, hormones, bactericides, illicit drugs, compounds such as caffeine and nicotine, and disinfection by-products [2,3]. ECs enter the environment through anthropogenic contamination. For instance, human beings consume large amounts of pharmaceutical and personal care products generating waste that often ends up in wastewater. Different compounds have been detected in municipal and natural water systems which are poured into through residential or commercial discharges [4]. For example, pharmaceutical and personal hygiene products generally present in human and animal excretions enter the environment through domestic wastewater via discharge from toilets, domestic water, among other sources [5]. Pharmaceutical wastewater is usually recalcitrant with high chemical oxygen demand (COD), high biological toxicity, low biodegradability, intense color, and unpleasant odor. It contains high concentrations of solvents, catalysts, additives, and reagents, especially antibiotics [6]. Wastewaters containing these compounds are discharged directly into the environment or are treated in wastewater treatment plants where ECs are not effectively eliminated due to their low concentrations and complexity [1,7,8]. One of the characteristics of these pollutants is that they do not have to be persistent in the environment to cause negative effects since their high transformation and elimination rates can be offset by their continuous introduction into the environment [9]. Generally, they are characterized by the following properties: high chemical stability, low biodegradability, high solubility in water, and low adsorption coefficient [10]. The emerging pollutants that cause the most concern are antibiotics and endocrine disruptors. Antibiotics such as sulfamethoxazole can induce bacterial resistance, even at low concentrations, through continuous exposure [11]. Endocrine disruptors such as estradiol and triclosan interfere with the endocrine system and disrupt the physiological function of hormones by mimicking, blocking, or disrupting their function thus affecting the health of humans and animal species [12–16]. Nicotine, sulfamethoxazole, triclosan and estradiol have been found in superficial water bodies such as rivers, lakes, seas, and oceans [1,17–22] and even in drinking water. Hence pointing out that the current treatment methods are not effective to eliminate these compounds [17–26].

Recently, the ability of some techniques to remove ECs from water, such as advanced oxidation and adsorption has been discussed. For instance, estrogens removal rates vary between 100 μgL^{-1} to 10 mgL^{-1} in time periods that oscillate between 7 to 300 min with electrochemical advanced oxidation treatment [27]; however, these processes present a high cost and difficulty of implementation [1]. On the other hand, adsorption methods have the disadvantage that they only generate a phase transfer. Nevertheless, they are still a valid option for removing ECs from water due to their simplicity and low cost. Among the different sorbents used for this purpose, clay minerals, such as bentonite and zeolite, have shown effective results as adsorbent and ion exchange media for water and wastewater treatment applications, especially for removing heavy metals, organic pollutants, and nutrients [28–30]. Bentonite and/or zeolite have been used to remove organic pollutants such as dyes, hormones, pharmaceuticals, caffeine, and other ECs [31–36]; the rate of adsorption in these materials depends on factors such as temperature and pH, as well as the chemical nature of the retained compound. The adsorption of estradiol in clay minerals is an exothermic and spontaneous process; levels of pH higher to 10 decrease the adsorption capability [37]. In contrast, nicotine adsorption is a spontaneous process, endothermic or exothermic depending on the adsorbent; nicotine is an electron donor due to the aliphatic nitrogen of the pyrrolidine ring. Moreover, pH determines the adsorption mechanism such as hydrogen bonds, π–π interaction, cation-π bonding, Van der Waals forces, inner-sphere complex formation, and electrostatic interactions [38].

Biologically-based filtration using aquatic invertebrates such as *Daphnia magna*, which live in biological-based wastewater treatment plants, has shown promising results to remove ECs [39], consequently becoming an optimal complement to adsorption processes. Besides, one of the main advantages of *D. magna* is their ability to reduce the biochemical oxygen demand in wastewater [40]. *D. magna* under direct solar radiation also could remove 80% of ECs combining biodegradation, photodegradation, adsorption, and adsorption processes [41]. Nonetheless, many of the evaluated studies do not consider how the presence of other contaminants affect the ECs removal performance and adsorption efficiency; for instance, studies used individual pollutants such as 17β-Estradiol, dyes, and phenol as model compounds for the adsorption studies [31,33–36].

The aim of this study is to investigate the removal of estradiol from wastewater using a combination of adsorption with Na-montmorillonite, Ca-montmorillonite, Zeolite, and biologically-based filtration with *Daphnia magna*. We employ an experimental design to assess the influence of adsorbent type and other EC's presence in estradiol's removal rate; nicotine, triclosan, and sulfamethoxazole were used for this purpose, as they are commonly found in water bodies along with estradiol.

2. Materials and Methods

2.1. Chemicals

Synthetic waters spiked with emerging contaminants were prepared for this study using Sigma Aldrich technical grade (99% purity) nicotine, triclosan, sulfamethoxazole, and estradiol. Table 1 reports the structures and properties of these four molecules [32,42–44]. For the mobile phase, high-performance liquid chromatography (HPLC) grade methanol was purchased from Merck, and Type 1 water (from a Milli-Q system, Merck Millipore) was used.

Table 1. Emerging contaminants targeted in the study.

Name/CAS	Description	Chemical Formula	Molar Mass gmol^{-1}	Water Solubility mgL^{-1}	pKa	Log Kow	Log Koc
17ß-Estradiol/50-28-2	Hormone	$C_{18}H_{24}O_2$	272.38	3.60, 27 °C	10.27	4.13	4.47
Nicotine/54-11-5	Stimulant	$C_{10}H_{14}N_2$	162.23	1,000,000, 25 °C	8.5	1.17	2
Triclosan/3380-34-5	Antibacterial	$C_{12}H_7Cl_3O_2$	289.54	10, 20 °C	7.9	4.76	3.54
Sulfamethoxazole/723-46-6	Antibiotic	$C_{10}H_{11}N_3O_3S$	253.3	610, 37 °C	1.6, 5.7	0.89	1.86

2.2. Bentonites and Zeolites

Ecuadorian natural Na-montmorillonite, Ca-montmorillonite and zeolites processed by the company Minmetec Ecuador Cia. Ltda. were used. The zeolite is a clinoptilolite, $(Ca)_3(Si_{30}Al_6)O_{72} \cdot H_2O$, belonging to the heulandite group. Na-montmorillonite, $(Na^{1+})_{0.33}(Al,Mg)_2(Si_4O_{10})(OH)_2 \cdot n(H_2O)$, is a smectite clay consisting mainly of montmorillonite, magnesium silicate, hydrated aluminum, and sodium. Ca-montmorillonite, $(Ca^{2+})_{0.33}(Al,Mg)_2(Si_4O_{10})(OH)_2 \cdot n(H_2O)$, is a smectite clay, consisting mainly of montmorillonite. Physicochemical characteristics are shown in Table 2 and chemical structure is presented in Figure 1.

2.3. Daphnia Magna Culture

The cultivation of *D. magna* was conducted according to Organization of Economic Co-operation and Development (OECD) 1981 standardized protocols. The photoperiod was set to a 14 h light:10 h dark cycle and the temperature was set at 20 ± 1 °C. Cultures were maintained in 100 mL of ASTM hard synthetic water and fed every three days with spirulina algae.

Table 2. Zeolite and montmorillonite composition and characteristics.

Composition	Zeolite	Ca-Montmorillonite	Na-Montmorillonite
Al_2O_3, %	16.87	13.06	17.59
SiO_2, %	63.78	60.64	60.50
Fe_2O_3, %	3.54	12.00	6.22
Na_2O, %	2.15	2.60	1.00
MgO, %	0.78	1.50	1.05
CaO, %	3.63	2.50	1.03
K_2O, %	2.45	0.5	1.25
pH	9.8	9	7
Max Humidity, %	7	7	10
Color	Brown—greenish	Light brown	White—creamy

Values taken of [42,45–47].

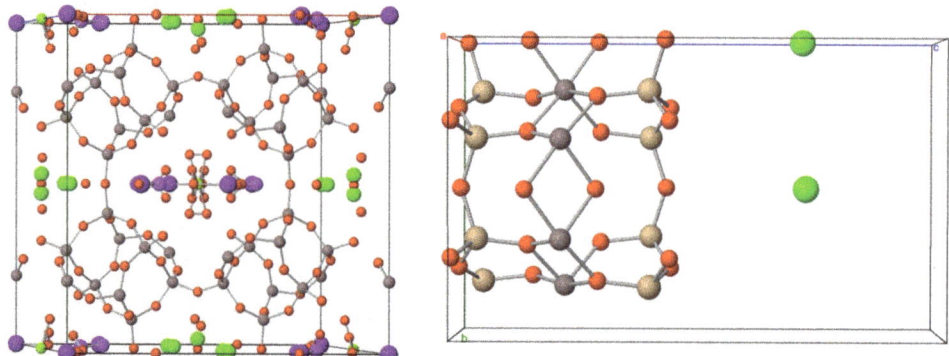

Figure 1. In the left Zeolite Clinoptilolite and in the right montmorillonite taken from mindat.org. Red ball: Oxygen, Light brown: Silica, Gray: Aluminum, Purple: Sodium or Potassium, Green: Sodium or Calcium.

2.4. Experimental Design

2.4.1. Zeolites and Bentonites Adsorption

Preliminary adsorption tests were performed to determine whether the clays could adsorb estradiol. Once the effectiveness of the materials to adsorb estradiol was verified, a Plackett–Burman experimental design was conducted; eight variables were selected to take into account: the concentration of estradiol, nicotine, sulfamethoxazole and triclosan, the mass of zeolite, Na-montmorillonite, Ca-montmorillonite, and the exposure time. For the assays, a stock solution of 100 mgL^{-1} of each of four compounds was prepared in a hydroalcoholic solution. Then, 15 mixtures of emerging contaminants were prepared with the concentrations shown in Table 3 and following the scheme depicted in Table 4. The aqueous mixtures were prepared with tap water up to a volume of 500 mL.

Table 3. Variables and levels of the applied experimental design.

Variable	Min (−1)	Medium (0)	Max (1)
Estradiol (mgL^{-1})	0.25	0.5	1
Nicotine (mgL^{-1})	0.25	0.5	1
Sulfamethoxazole (mgL^{-1})	0.25	0.5	1
Triclosan (mgL^{-1})	0.25	0.5	1
Na-montmorillonite (g)	2	4	8
Ca-montmorillonite (g)	2	4	8
Zeolita (g)	2	4	8
Time (minutes)	15	30	60

All the experiments were performed in fluid bed batch reactors. The concentration measurements of the four studied ECs were made by HPLC-DAD to assess the efficiency of the suggested removal method before and after the use of the clay and the *D. magna* treatments. In the end of the experiment, samples were centrifuged and filtered before the HPLC analysis (Section 2.5). The existing difference of the same sample, expressed as a percentage, was used as the response variable of the experimental design. The software used for the data analysis was MINITAB 17.

Table 4. Variable levels for each trial as applied in the Plackett–Burman design.

Trial	Estradiol	Nicotine	Sulfamethoxazole	Triclosan	Na-Montmorillonite	Ca-Montmorillonite	Zeolite	Time
1	1	−1	1	−1	−1	−1	1	1
2	1	1	−1	1	−1	−1	−1	1
3	−1	1	1	−1	1	−1	−1	−1
4	1	−1	1	1	−1	1	−1	−1
5	1	1	−1	1	1	−1	1	−1
6	1	1	1	−1	1	1	−1	1
7	−1	1	1	1	−1	1	1	−1
8	−1	−1	1	1	1	−1	1	1
9	−1	−1	−1	1	1	1	−1	1
10	1	−1	−1	−1	1	1	1	−1
11	−1	1	−1	−1	−1	1	1	1
12	−1	−1	−1	−1	−1	−1	−1	−1
13	0	0	0	0	0	0	0	0
14	0	0	0	0	0	0	0	0
15	0	0	0	0	0	0	0	0

2.4.2. *Daphnia magna* Adsorption

Approximately 50 *D. magna* individuals were kept in an aliquot of 100 mL of water samples for 8 days; all assays were performed in three replicates. The experimental setup was kept at room temperature, 20 °C at 12 h of light and 12 h of darkness. *D. magna* were fed with 1 mL of a mixture of 12 mg of yeast and 30 mg of spirulina every 2 days. After the adsorption experiments, aliquots of 100 mL of water samples were centrifuged (Eppendorf Centrifuge 5702R) at 4000 rpm for 15 min. The supernatant was filtered using 0.45 µm PVDF filters. Then, the estradiol concentration was determined by liquid chromatography to calculate the removal percentage achieved.

2.5. HPLC Analysis

RP-HPLC analyses were performed using means of an HPLC-DAD instrument (Thermo Scientific UltiMate 3000, USA). An isocratic elution program was used with a mixture of 60:40 methanol:water v/v as mobile phase, and a C18 column as stationary phase. The flow rate was 0.6 mLs^{-1}, while the column was thermostated at 25 °C. The injection volume was 5 µL for all standards and samples. Compound elution was detected at 220 nm. Under these conditions, estradiol retention time was 6.62 min. The LOD and LOQ were respectively: estradiol: 0.540 µgL^{-1} and 1.042 µgL^{-1}; nicotine: 11.55 µgL^{-1} and 22.29 µgL^{-1}; triclosan: 0.1645 µgL^{-1} and 0.603 µgL^{-1}; sulfamethoxazole 4.89 µgL^{-1}, and 9.44 µgL^{-1}.

2.6. Isoterm

Adsorption isotherms are used to evaluate the equilibrium of particles in a system with a liquid and a solid phase, at a constant temperature. For the development of this study, Langmuir and Freundlich models were used to calculate the adsorption isotherms of estradiol on the different clays used [48,49]. The Langmuir isotherm was developed from the assumption that adsorbate molecules form a monolayer on the surface of the adsorbent. Considering that the adsorbed molecules do not interact with each other, it is assumed that the adsorption of adsorbate at a specific site is independent of what happens with neighboring sites. The Langmuir isotherm is represented by the Equation (1) where C_e is the adsorbate concentration at equilibrium, q_e is the amount of adsorbate per unit mass of adsorbent at equilibrium, q_m is the maximum amount of adsorbate adsorbed per unit mass of adsorbent for the formation of the complete monolayer on the surface of the adsorbent, K_L is the Langmuir constant related to the adsorption energy [50,51]. The Freundlich isotherm, Equation (2), is applied when adsorption processes occur on heterogeneous surfaces, thus being a model that can explain both monolayer and multilayer adsorption. The expression resulting from this isotherm defines the heterogeneity of the surface and how the active sites and their energies are exponentially distributed [52,53]. K_F is the Freundlich constant that is related to the adsorption capacity; n represents the heterogeneity factor and $1/n$ is related to the adsorption intensity.

$$q_e = \frac{q_m K_L C_e}{1 + K_L C_e} \quad (1)$$

$$sq_e = K_L C_e^{\frac{1}{n}} \quad (2)$$

The parameters of the isotherms were obtained from measuring the estradiol concentration in the supernatant of a suspension of 50 mL of estradiol solutions at different concentrations, in contact with 100 mg of each of the clays. It was left overnight at a constant temperature of 30 °C. The concentration of the estradiol standards used were: 0.5, 1.0, 2.5, 3, 3.5, 4, 4.5, 5, 10, 15, and 20 mgL^{-1}. Absorbance was measured using a UV-Vis Spectrophotometer (Thermo Scientific, Evolution 60) at a wavelength of 280 nm. Measurements were made in duplicate.

3. Results and Discussion

The results of the set of experiments are presented in Table 5. In the adsorption tests with clays, the removal of estradiol was quantified, the removal percentage was from 40 to 92% (0.1 –0.92 mgL^{-1}, respectively). Nicotine, sulfamethoxazole, and triclosan were not quantified in this study. Other studies show that these compounds are poorly or not removed with natural clays and a pre-treatment of clay is necessary to achieve the adsorption [54–56]. Considering the soil adsorption coefficient (Koc) which measures the amount of chemical substance adsorbed onto soil per amount of water, estradiol is prone to be adsorbed in clays.

Table 5. Amount and percentage of estradiol removal obtained in each applied treatment.

Trial	Estradiol Removal, (mg/%)		
	Clays	D. magna	Total
1	0.10/40.00	0.09/61.94	0.19/77.16
2	0.92/92.29	0.003/3.26	0.93/92.54
3	0.76/76.36	0.22/93.08	0.98/98.36
4	0.16/64.32	0.08/91.94	0.24/97.12
5	0.71/71.03	0.21/71.78	0.92/91.82
6	0.57/56.51	0.35/79.47	0.91/91.07
7	0.85/85.17	0.09/62.54	0.94/94.44
8	0.09/37.21	0.00/0.00	0.09/37.21
9	0.13/50.15	0.10/83.98	0.23/92.01
10	0.11/44.53	0.12/87.45	0.23/93.04
11	0.69/68.73	0.27/87.22	0.96/96.00
12	0.14/56.50	0.07/60.85	0.21/82.97
13	0.32/63.74	0.16/90.10	0.48/96.41
14	0.34/67.06	0.18/89.62	0.48/96.58
15	0.31/61.79	0.11/57.63	0.42/83.81

The experiments with natural clays showed the lowest removal with trials 1, 8, 9, 10, and 12 which achieved a removal of estradiol of 0.10, 0.09, 0.13, 0.11, and 0.14 mgL^{-1} respectively. All these assays were performed with nicotine concentration in the lowest level (0.25 mgL^{-1}). On the other hand, the experiments with natural clays where greater estradiol removal was achieved were trials number 2, 3, 5, 7, and 11, corresponding to 0.92, 0.76, 0.71, 0.85, and 0.69 mgL^{-1}, respectively. In these cases, all the experiments contain the maximum amount of nicotine.

The results of the analysis of the experimental design in clays are presented in Figure 2. The significant effects are shown in a red square, then, the significant factors in the experimental design with a confidence level of 95% are nicotine and triclosan as a positive influence in the estradiol adsorption, and Na-montmorillonite, Zeolite, and time with negative influence in the estradiol adsorption. In the Pareto chart, the five factors are presented where nicotine is the most important factor. The level of significance of each factor can be seen in Table 6.

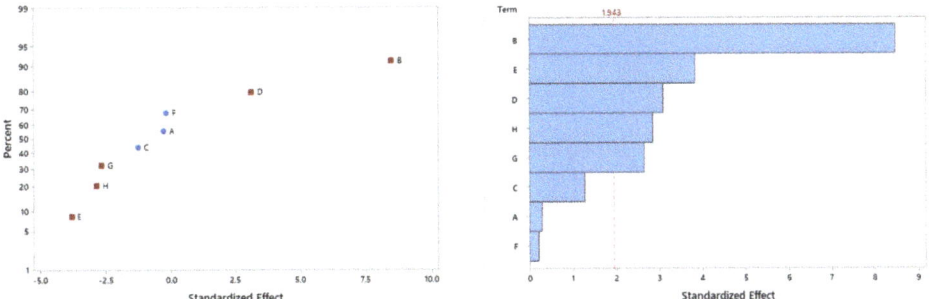

Figure 2. Standardized effect in a normal graph and a Pareto chart of the factors in estradiol removal. A: estradiol, B: nicotine, C: sulfamethoxazole, D: triclosan, E: Na-montmorillonite, F: Ca-montmorillonite, G: Zeolite, and H: Time.

Table 6. ANOVA of the experimental design performed to assess estradiol removal with clays.

Source	DF	Adj SS	Adj MS	F-Value	p-Value
Model	8	3251.05	406.38	14.12	0.002 **
Linear	8	3251.05	406.38	14.12	0.002 **
Estradiol	1	2.47	2.47	0.09	0.780
Nicotine	1	2063.84	2063.84	71.69	0.000 ****
Sulfamethoxazole	1	46.73	46.73	1.62	0.250
Triclosan	1	275.86	275.86	9.58	0.021 *
Na-montmorillonite	1	422.67	422.67	14.68	0.009 **
Ca-montmorillonite	1	1.32	1.32	0.05	0.838
Zeolite	1	203.86	203.86	7.08	0.037 *
Time	1	234.32	234.32	8.14	0.029 *
Error	6	172.73	28.79		
Curvature	1	12.65	12.65	0.40	0.557
Lack-of-Fit	3	145.91	48.64	6.87	0.130
Pure Error	2	14.17	7.08		
Total	14	3423.79			

Adj SS: Adjusted sums of squares; Adj MS: Adjusted mean squares; Significance at $p < (0.05)$ *, $p < (0.01)$ **, $p < (0.0000)$ ****.

The model obtained is statistically significant ($p(0.02) < 0.05$, Table 6) and this is presented in Equation (3). The effects of each factor are presented in Figure 3. Factors that reduced the adsorption of estradiol were sulfamethoxazole, Ca-montmorillonite, zeolite, and long exposure. In contrast, the nicotine and triclosan presence increases the adsorption of estradiol.

$$\% \text{ Estradiol removal} = 62.4 - 0.5*\text{Estradiol} + 13.1*\text{Nicotine} - 2*\text{Sulfamethoxazole} + 4.8*\text{Triclosan} - 5.9*\text{Na-montmorillonite} - 0.3*\text{Ca-montmorillonite} - 4.1*\text{Zeolite} - 4.4*\text{Time} \quad (3)$$

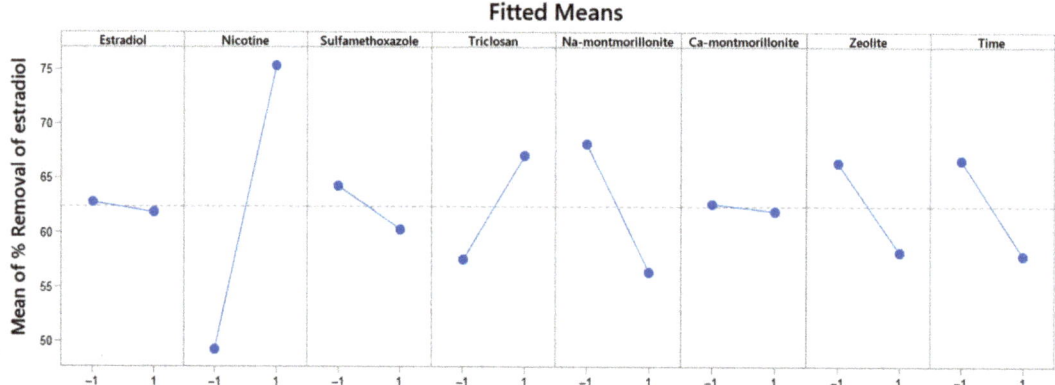

Figure 3. Fitted means effect of each factor involved in estradiol removal.

The fitted means effects graph (Figure 3) reveals that the initial concentration of estradiol does not make a difference in terms of the final percentage of estradiol removal. Although a lower concentration slightly improves the removal efficiency, it is not statistically significant. It was expected that a higher concentration drives removal, given that the adsorption kinetics for this compound corresponds to a second-order reaction [57]. This result is explained by the short range and low concentrations used in this study (1 to 0.25 mgL^{-1}), which were selected to be similar to those amounts found in the environment [1]. Furthermore, shorter elapsed time favors adsorption; this result is consistent with a pseudo second order kinetics, meaning that the adsorption occurs early while the concentration is high enough but when concentration decreases, adsorption decreases too, and equilibrium is achieved. Similar results were found in other studies [58,59].

The results show that nicotine and triclosan are statistically significant to promote the estradiol adsorption; possibly, they stimulate the estradiol removal due to the formation of a complex between the three compounds. Estradiol can act as a two hydrogen bond donor and two hydrogen bond acceptor. Triclosan is a hydrogen bond donor and two hydrogen bond acceptor. Nicotine is an acceptor of two hydrogen bonds [42]. These characteristics for forming bonds favor the formation of complexes between the three compounds. Nicotine and triclosan, which are located in the end of estradiol, form hydrogen bonds with their OH groups [60–62]. The formation of the complex does not affect its binding with the adsorbent, since Van der Waals forces are the main interaction with the adsorbent. Van der Waals bonds increase as the length of the nonpolar part of the complex increases [63,64]. In turn, this ability to form a complex can facilitate multilayer adsorption.

Figure 4 shows the adsorption isotherms of estradiol on the three clays used in this study. Langmuir's model was used to describe the adsorption equilibrium, hypothesizing the existence of a monolayer adsorption, and the linearized form of the Langmuir model was used to obtain the parameters of the model. The Freundlich model was calculated to describe the adsorption effect of the multilayer, for which the linearized form of the model was used. The Freundlich model parameters were optimized using the algorithm Generalized Reduced Gradient (GRG) Nonlinear. Parameters of the Langmuir and Freundlich models are presented in Table 7. Although the R^2 values of the linearized Langmuir models are higher than the optimized Freundlich models, a strong leverage effect can be observed in the upper part of the curve, so the Langmuir model overestimates the effect of the monolayer. On the contrary, the optimized Freundlich model has a greater similarity with the curve described by the experimental data.

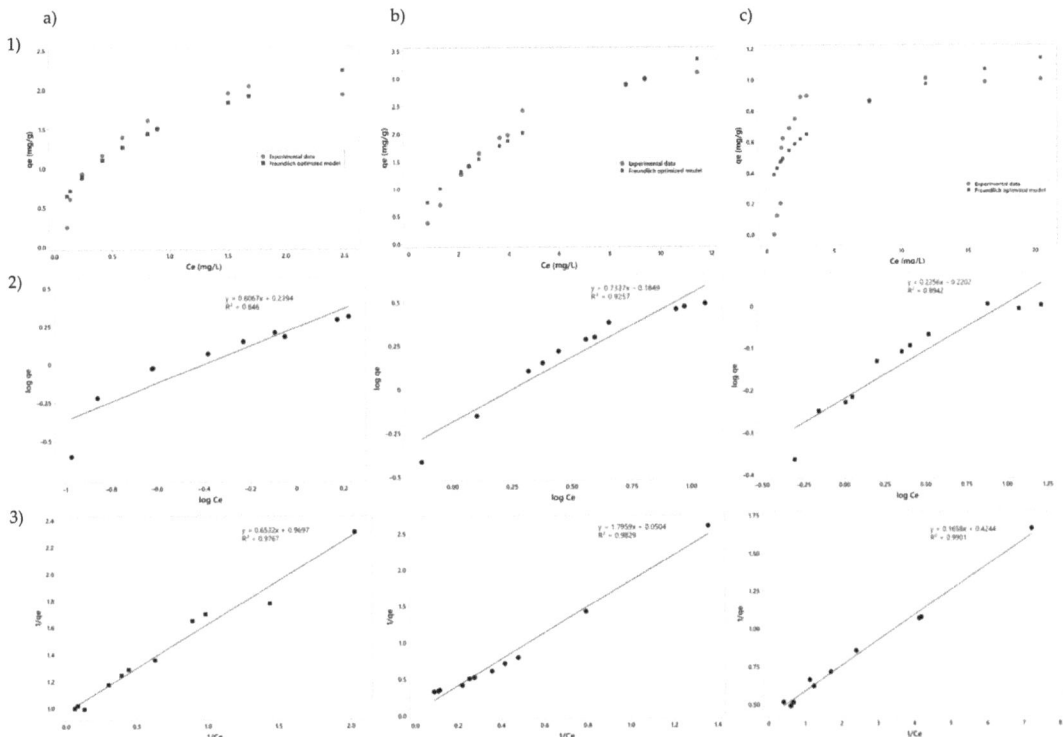

Figure 4. Adsorption isotherm for (**a**) Na-montmorillonite, (**b**) Ca-montmorillonite, and (**c**) Zeolite; (1) Experimental data, (2) the Freundlich adsorption isotherms, and (3) the Langmuir adsorption isotherms. qe: number of milligrams of adsorbate that is adsorbed per gram of adsorbent; Ce: concentration of adsorbate in solution when equilibrium has been reached.

Table 7. Langmuir and Freundlich isotherm models parameters for clays.

Isotherm	Model Parameters	Na-Montmorillonite	Ca-Montmorillonite	Zeolite
Langmuir	qmax (mg g^{-1})	2.35	19.84	1.031
	KL	0.256	35.63	1.4845
	R^2	0.990	0.983	0.977
Freundlich	KF	1.565	0.880	0.478
	n	2.531	1.828	3.485
	R^2	0.972	0.989	0.967

The results show that natural bentonites montmorillonites and zeolites effectively removed estradiol despite the presence of other contaminants. Nevertheless, the combination of the clays did not represent an improvement in the removal system. The clays for this study did not receive any previous treatment since the intention is to evaluate their adsorbent capacity under natural conditions for their use in wastewater treatments. The value of the constant K_L of the Langmuir model shows that Ca-montmorillonite has a higher binding force for estradiol than the other two clays. In addition, for this reason, the value of q_{max} is the highest of all. In the case of the Freundlich model, Ca-montmorillonite has a value of n, lower than the other two clays, so it should have a higher intensity of adsorption, saturating more slowly. When comparing the results obtained from the clay isotherms and the removal results from the experimental design, the interaction between different molecules changes their ability to be adsorbed. In this specific case, estradiol would form a complex with triclosan and nicotine, and its preference for Ca-montmorillonite changes to the point that Ca-montmorillonite does not have a significant effect on the experimental design. This could be due to the tendency of the complex to be adsorbed, not in a monolayer, but in a multilayer, and the amount of Ca-montmorillonite is sufficient for this purpose. According to Figure 2, a graph of standardized effects, it can be observed that all clays are aligned with the non-significant variables, thus it can be considered that the effect of the three clays on estradiol removal was non-significant. Consequently, nicotine and triclosan would be the variables influencing estradiol removal in this system.

Regarding the use of the *Daphnia magna* as a natural filter, the estradiol removal ranges between 0–93%. The trials where low estradiol concentration was removed were number 2 and 8 (Table 5) corresponding to 0 mgL^{-1} in both cases; their common characteristic is the presence of triclosan concentration (1 mgL^{-1}). This result is explained since triclosan can bioaccumulate via the food chain causing adverse effects depending on the concentration [63], so it might interfere with the ability of *D. magna* to metabolize or adsorb estradiol. The maximum removal of estradiol was obtained in trials 3, 6, and 11 corresponding to 0.22, 0.35, and 0.27 mgL^{-1}. In all these trials, nicotine was in the maximum concentration, whereas triclosan was in the minimum concentration. The most feasible route for *D. magna* to remove these pollutants is biosorption and, secondly, ingestion; in both cases, the compounds/metabolites are subsequently eliminated from the *D. magna* through their excretions, growth, breeding, and desorption [65–67]. The formation of the complex not only improved the adsorption of zeolites or montmorillonites, but also favored *Daphnia magna*, since their mortality does not occur at the estradiol and nicotine concentrations studied.

Overall, the best conditions for estradiol removal were those in trials 3, 4, 11, 13, and 14 where 0.93, 0.98, 0.96, 0.48, and 0.48 mgL^{-1} were eliminated, respectively. Due to the kinetics of the system, the tests with the least amount of estradiol also showed the lowest removal. The environmental concentration of estradiol in surface waters is in the range of nanograms per liter, the results of this study show the feasible use of the proposed combination of physicochemical and biological treatments for removing estradiol of water. However, more studies should be done to evaluate the interaction with other ECs.

Author Contributions: Conceptualization, P.T., V.P.-V. and A.A.; methodology, A.P.-G. and V.P.-V.; software, P.T.; validation, E.J.-N. and I.C.-A.; investigation, A.P.-G. and V.P.-V.; writing—original draft preparation, V.P.-V., A.P.-G., E.J.-N. and I.C.-A.; writing—review and editing, M.C., V.P.-V., A.P.-G., I.C.-A., E.J.-N. and J.F.C. All authors have read and agreed to the published version of the manuscript.

Funding: This research was funded by CEDIA, grant number CEPRA-XIV-2020-09.

Institutional Review Board Statement: Not applicable.

Informed Consent Statement: Not applicable.

Acknowledgments: The authors would like to thank the Corporación Ecuatoriana para el Desarrollo de la Investigación y Academia—CEDIA for their contribution in innovation, through the CEPRA projects, especially the project CEPRA-XIV-2020-09 "Determinación del impacto y ocurrencia de Contaminantes Emergentes en ríos de la Costa Ecuatoriana y propuestas de tratamiento para su remoción".

Conflicts of Interest: The authors declare no conflict of interest.

References

1. Vélez, V.P.P.; Esquivel-Hernández, G.; Cipriani-Avila, I.; Mora-Abril, E.; Cisneros, J.F.; Alvarado, A.; Abril-Ulloa, V. Emerging contaminants in trans-American waters. *Ambient. Agua Interdiscip. J. Appl. Sci.* **2019**, *14*, 1–26. [CrossRef]
2. Capparelli, M.V.; Cipriani-Avila, I.; Jara-Negrete, E.; Acosta-López, S.; Acosta, B.; Pérez-González, A.; Molinero, J.; Pinos-Vélez, V. Emerging contaminants in the northeast Andean foothills of Amazonia: The case of study of the city of Tena, Napo, Ecuador. *Bull. Environ. Contam. Toxicol.* **2021**, 1–9. [CrossRef]
3. Richardson, S.D.; Ternes, T.A. Water analysis: Emerging contaminants and current issues. *Anal. Chem.* **2011**, *83*, 4614–4648. [CrossRef] [PubMed]

4. Abu Hasan, H.; Abdullah, S.R.S.; Alattabi, A.W.; Nash, D.A.H.; Anuar, N.; Rahman, N.A.; Titah, H.S. Removal of ibuprofen, ketoprofen, COD and nitrogen compounds from pharmaceutical wastewater using aerobic suspension-sequencing batch reactor (ASSBR). *Sep. Purif. Technol.* **2016**, *157*, 215–221. [CrossRef]
5. Bell, K.Y.; Wells, M.J.; Traexler, K.A.; Pellegrin, M.-L.; Morse, A.; Bandy, J. Emerging pollutants. *Water Environ. Res.* **2011**, *83*, 1906–1984. [CrossRef]
6. Huang, B.; Wang, H.-C.; Cui, D.; Zhang, B.; Chen, Z.-B.; Wang, A.-J. Treatment of pharmaceutical wastewater containing β-lactams antibiotics by a pilot-scale anaerobic membrane bioreactor (AnMBR). *Chem. Eng. J.* **2018**, *341*, 238–247. [CrossRef]
7. Gil, M.; Soto, A.; Usma, J.; Gutierrez, O. Contaminantes emergentes en aguas, efectos y posibles tratamientos. *Prod. Limpia* **2012**, *7*, 52–73.
8. Meffe, R.; de Bustamante, I. Emerging organic contaminants in surface water and groundwater: A first overview of the situation in Italy. *Sci. Total Environ.* **2014**, *481*, 280–295. [CrossRef] [PubMed]
9. Becerril, J. Contaminantes emergentes en el agua. *Rev. Digit. Univ.* **2009**, *10*, 1067–6079.
10. Zwiener, C. Occurrence and analysis of pharmaceuticals and their transformation products in drinking water treatment. *Anal. Bioanal. Chem.* **2006**, *387*, 1159–1162. [CrossRef]
11. Hernandez, F.; Sancho, J.V.; Ibañez, M.; Guerrero, C. Antibiotic residue determination in environmental waters by LC-MS. *TrAC Trends Anal. Chem.* **2007**, *26*, 466–485. [CrossRef]
12. Aufartová, J.; Mahugo-Santana, C.; Sosa-Ferrera, Z.; Santana-Rodríguez, J.J.; Nováková, L.; Solich, P. Determination of steroid hormones in biological and environmental samples using green microextraction techniques: An overview. *Anal. Chim. Acta* **2011**, *704*, 33–46. [CrossRef] [PubMed]
13. Estrada-Arriaga, E.B.; Cortés-Muñoz, J.E.; González-Herrera, A.; Calderón-Mólgora, C.G.; Rivera-Huerta, M.D.L.; Ramírez-Camperos, E.; Montellano-Palacios, L.; Gelover-Santiago, S.L.; Pérez-Castrejón, S.; Cardoso-Vigueros, L.; et al. Assessment of full-scale biological nutrient removal systems upgraded with physico-chemical processes for the removal of emerging pollutants present in wastewaters from Mexico. *Sci. Total Environ.* **2016**, *571*, 1172–1182. [CrossRef] [PubMed]
14. Huerta-Fontela, M.; Galceran, M.T.; Ventura, F. Fast liquid chromatography—Quadrupole-linear ion trap mass spectrometry for the analysis of pharmaceuticals and hormones in water resources. *J. Chromatogr. A* **2010**, *1217*, 4212–4222. [CrossRef]
15. Niemuth, N.J.; Klaper, R.D. Emerging wastewater contaminant metformin causes intersex and reduced fecundity in fish. *Chemosphere* **2015**, *135*, 38–45. [CrossRef]
16. Siddique, S.; Kubwabo, C.; Harris, S. A review of the role of emerging environmental contaminants in the development of breast cancer in women. *Emerg. Contam.* **2016**, *2*, 204–219. [CrossRef]
17. Campanha, M.B.; Awan, A.; de Sousa, D.N.R.; Grosseli, G.M.; Mozeto, A.A.; Fadini, P.S. A 3-year study on occurrence of emerging contaminants in an urban stream of São Paulo State of Southeast Brazil. *Environ. Sci. Pollut. Res.* **2015**, *22*, 7936–7947. [CrossRef]
18. Fossi, M.C.; Baini, M.; Panti, C.; Galli, M.; Jiménez, B.; Muñoz-Arnanz, J.; Marsili, L.; Finoia, M.G.; Ramírez-Macías, D. Are whale sharks exposed to persistent organic pollutants and plastic pollution in the Gulf of California (Mexico)? First ecotoxicological investigation using skin biopsies. *Comp. Biochem. Physiol. Part C Toxicol. Pharmacol.* **2017**, *199*, 48–58. [CrossRef] [PubMed]
19. Combi, T.; Pintado-Herrera, M.G.; Lara-Martin, P.A.; Miserocchi, S.; Langone, L.; Guerra, R. Distribution and fate of legacy and emerging contaminants along the Adriatic Sea: A comparative study. *Environ. Pollut.* **2016**, *218*, 1055–1064. [CrossRef]
20. Roberts, J.; Kumar, A.; Du, J.; Hepplewhite, C.; Ellis, D.J.; Christy, A.G.; Beavis, S.G. Pharmaceuticals and personal care products (PPCPs) in Australia's largest inland sewage treatment plant, and its contribution to a major Australian river during high and low flow. *Sci. Total Environ.* **2016**, *541*, 1625–1637. [CrossRef]
21. Voloshenko-Rossin, A.; Gasser, G.; Cohen, K.; Gun, J.; Cumbal, L.; Parra-Morales, W.; Sarabia, F.; Ojeda, F.; Lev, O. Emerging pollutants in the Esmeraldas watershed in Ecuador: Discharge and attenuation of emerging organic pollutants along the San Pedro—Guayllabamba—Esmeraldas rivers. *Environ. Sci. Process. Impacts* **2014**, *17*, 41–53. [CrossRef]
22. Bai, X.; Lutz, A.; Carroll, R.; Keteles, K.; Dahlin, K.; Murphy, M.; Nguyen, D. Occurrence, distribution, and seasonality of emerging contaminants in urban watersheds. *Chemosphere* **2018**, *200*, 133–142. [CrossRef] [PubMed]
23. Bolong, N.; Ismail, A.F.; Salim, M.R.; Matsuura, T. A review of the effects of emerging contaminants in wastewater and options for their removal. *Desalination* **2009**, *239*, 229–246. [CrossRef]
24. Tejada, C.; Quiñones, E.; Peña, M. Contaminantes emergentes en aguas: Metabolitos de fármacos. Una Revision. *Rev. Fac. de Cienc. Básicas* **2014**, *10*, 80–101. [CrossRef]
25. García, C.; Gortáres, P.; Drogui, P. Contaminantes emergentes: Efectos y tratamientos de remoción. *Rev. Química Viva* **2011**, *10*, 96–105.
26. Batt, A.L.; Furlong, E.T.; Mash, H.E.; Glassmeyer, S.T.; Kolpin, D.W. The importance of quality control in validating concentrations of contaminants of emerging concern in source and treated drinking water samples. *Sci. Total Environ.* **2017**, *579*, 1618–1628. [CrossRef] [PubMed]
27. Torres, N.H.; Santos, G.D.O.S.; Ferreira, L.F.R.; Américo-Pinheiro, J.H.P.; Eguiluz, K.I.B.; Salazar-Banda, G.R. Environmental aspects of hormones estriol, 17β-estradiol and 17α-ethinylestradiol: Electrochemical processes as next-generation technologies for their removal in water matrices. *Chemosphere* **2021**, *267*, 128888. [CrossRef] [PubMed]
28. Abdullah, R.; Abustan, I.; Ibrahim, A.N.M. Wastewater treatment using bentonite, the combinations of bentonite-zeolite, bentonite-alum, and bentonite-limestone as adsorbent and coagulant. *Int. J. Environ. Sci.* **2013**, *4*, 379.

29. Delkash, M.; Bakhshayesh, B.E.; Kazemian, H. Using zeolitic adsorbents to cleanup special wastewater streams: A review. *Microporous Mesoporous Mater.* **2015**, *214*, 224–241. [CrossRef]
30. Yuna, Z. Review of the natural, modified, and synthetic zeolites for heavy metals removal from wastewater. *Environ. Eng. Sci.* **2016**, *33*, 443–454. [CrossRef]
31. Mirzaei, N.; Hadi, M.; Gholami, M.; Fard, R.F.; Aminabad, M.S. Sorption of acid dye by surfactant modificated natural zeolites. *J. Taiwan Inst. Chem. Eng.* **2016**, *59*, 186–194. [CrossRef]
32. Rossner, A.; Snyder, S.; Knappe, D.R. Removal of emerging contaminants of concern by alternative adsorbents. *Water Res.* **2009**, *43*, 3787–3796. [CrossRef]
33. Shi, J.; Yang, Z.; Dai, H.; Lu, X.; Peng, L.; Tan, X.; Shi, L.; Fahim, R. Preparation and application of modified zeolites as adsorbents in wastewater treatment. *Water Sci. Technol.* **2018**, *2017*, 621–635. [CrossRef]
34. Wang, S.; Peng, Y. Natural zeolites as effective adsorbents in water and wastewater treatment. *Chem. Eng. J.* **2010**, *156*, 11–24. [CrossRef]
35. Yousef, R.I.; El-Eswed, B.; Al-Muhtaseb, A.H. Adsorption characteristics of natural zeolites as solid adsorbents for phenol removal from aqueous solutions: Kinetics, mechanism, and thermodynamics studies. *Chem. Eng. J.* **2011**, *171*, 1143–1149. [CrossRef]
36. Toor, M.; Jin, B.; Dai, S.; Vimonses, V. Activating natural bentonite as a cost-effective adsorbent for removal of Congo-red in wastewater. *J. Ind. Eng. Chem.* **2015**, *21*, 653–661. [CrossRef]
37. Tong, X.; Li, Y.; Zhang, F.; Chen, X.; Zhao, Y.; Hu, B.; Zhang, X. Adsorption of 17β-estradiol onto humic-mineral complexes and effects of temperature, pH, and bisphenol A on the adsorption process. *Environ. Pollut.* **2019**, *254*, 112924. [CrossRef] [PubMed]
38. Anastopoulos, I.; Pashalidis, I.; Orfanos, A.G.; Manariotis, I.D.; Tatarchuk, T.; Sellaoui, L.; Bonilla-Petriciolet, A.; Mittal, A.; Núñez-Delgado, A. Removal of caffeine, nicotine and amoxicillin from (waste)waters by various adsorbents. A review. *J. Environ. Manag.* **2020**, *261*, 110236. [CrossRef]
39. Matamoros, V.; Sala, L.; Salvado, V. Evaluation of a biologically-based filtration water reclamation plant for removing emerging contaminants: A pilot plant study. *Bioresour. Technol.* **2012**, *104*, 243–249. [CrossRef]
40. Shiny, K. Biotreatment of wastewater using aquatic invertebrates, *Daphnia magna* and *Paramecium caudatum*. *Bioresour. Technol.* **2005**, *96*, 55–58. [CrossRef] [PubMed]
41. Ortiz, G.M. Preparación y Actividad Catalítica de Sistemas Cromo-Arcilla y Níquel-Arcilla. Ph.D. Thesis, Universidad de Salamanca, Salamanca, Spain, 2012. Available online: http://purl.org/dc/dcmitype/Text, (accessed on 15 May 2021).
42. National Center for Biotechnology Information. Available online: https://pubchem.ncbi.nlm.nih.gov (accessed on 10 March 2021).
43. Barron, L.; Havel, J.; Purcell, M.; Szpak, M.; Kelleher, B.; Paull, B. Predicting sorption of pharmaceuticals and personal care products onto soil and digested sludge using artificial neural networks. *Analyst* **2009**, *134*, 663–670. [CrossRef]
44. Karnjanapiboonwong, A.; Morse, A.N.; Maul, J.D.; Anderson, T. Sorption of estrogens, triclosan, and caffeine in a sandy loam and a silt loam soil. *J. Soils Sediments* **2010**, *10*, 1300–1307. [CrossRef]
45. Minmetec. *Ficha Técnica Agrolite*; Minmetec: Cuenca, Ecuador, 2020.
46. Minmetec. *Ficha Técnica Bentonita Sódica*; Minmetec: Cuenca, Ecuador, 2019.
47. Minmetec. *Ficha Técnica Bentonita Cálcica*; Minmetec: Cuenca, Ecuador, 2021.
48. Yuh-Shan, H. Citation review of Lagergren kinetic rate equation on adsorption reactions. *Scientometrics* **2004**, *59*, 171–177. [CrossRef]
49. Wang, J.; Guo, X. Adsorption isotherm models: Classification, physical meaning, application and solving method. *Chemosphere* **2020**, *258*, 127279. [CrossRef] [PubMed]
50. Toor, M.; Jin, B. Adsorption characteristics, isotherm, kinetics, and diffusion of modified natural bentonite for removing diazo dye. *Chem. Eng. J.* **2012**, *187*, 79–88. [CrossRef]
51. Gallouze, H.; Akretche, D.-E.; Daniel, C.; Coelhoso, I.; Crespo, J.G. Removal of synthetic estrogen from water by adsorption on modified bentonites. *Environ. Eng. Sci.* **2021**, *38*, 4–14. [CrossRef]
52. Yang, C.-H. Statistical mechanical study on the Freundlich isotherm equation. *J. Colloid Interface Sci.* **1998**, *208*, 379–387. [CrossRef] [PubMed]
53. Boparai, H.K.; Joseph, M.; O'Carroll, D. Kinetics and thermodynamics of cadmium ion removal by adsorption onto nano zerovalent iron particles. *J. Hazard. Mater.* **2011**, *186*, 458–465. [CrossRef]
54. De Rezende, J.C.T.; Ramos, V.H.S.; Silva, A.S.; Santos, E.; Oliveira, H.A.; de Jesus, E. Assessment of sulfamethoxazole adsorption capacity on Pirangi clay from the state of Sergipe, Brazil, modified by heating and addition of organic cation. *Cerâmica* **2019**, *65*, 626–634. [CrossRef]
55. Styszko, K.; Nosek, K.; Motak, M.; Bester, K. Preliminary selection of clay minerals for the removal of pharmaceuticals, bisphenol A and triclosan in acidic and neutral aqueous solutions. *Comptes Rendus Chim.* **2015**, *18*, 1134–1142. [CrossRef]
56. Behera, S.K.; Oh, S.-Y.; Park, H.-S. Sorption of triclosan onto activated carbon, kaolinite and montmorillonite: Effects of pH, ionic strength, and humic acid. *J. Hazard. Mater.* **2010**, *179*, 684–691. [CrossRef] [PubMed]
57. Parsegian, V.A. *Van der Waals Forces: A Handbook for Biologists, Chemists, Engineers, and Physicists*; Cambridge University Press: Cambridge, UK, 2005.
58. Liu, J.; Carr, S.A. Removal of estrogenic compounds from aqueous solutions using zeolites. *Water Environ. Res.* **2013**, *85*, 2157–2163. [CrossRef]
59. Singhal, J.; Singh, R. Studies on the adsorption of nicotine on kaolinites. *Soil Sci. Plant Nutr.* **1976**, *22*, 35–41. [CrossRef]

60. Grabowski, S.J.; Leszczynski, J. Unrevealing the nature of hydrogen bonds: π-electron delocalization shapes H-bond features. Intramolecular and intermolecular resonance-assisted hydrogen bonds. In *Hydrogen Bonding—New Insights*; Grabowski, S.J., Ed.; Springer: Dordrecht, The Netherlands, 2006; pp. 487–512. ISBN 9781402048531.
61. Shikii, K.; Seki, H.; Sakamoto, S.; Sei, Y.; Utsumi, H.; Yamaguchi, K. Intermolecular hydrogen bonding of steroid compounds: PFG NMR diffusion study, cold-spray ionization (CSI)-MS and X-ray analysis. *Chem. Pharm. Bull.* **2005**, *53*, 792–795. [CrossRef] [PubMed]
62. Leermakers, F.; Eriksson, J.C.; Lyklema, H. Chapter 4—Association colloids and their equilibrium modelling. In *Fundamentals of Interface and Colloid Science*; Lyklema, J., Ed.; Soft Colloids; Elsevier: Amsterdam, The Netherlands, 2005; Volume 5, pp. 4.1–4.123.
63. Thanhmingliana, T.; Lalhriatpuia, C.; Tiwari, D.; Lee, S.-M. Efficient removal of 17β-estradiol using hybrid clay materials: Batch and column studies. *Environ. Eng. Res.* **2016**, *21*, 203–210. [CrossRef]
64. Peng, Y.; Luo, Y.; Nie, X.-P.; Liao, W.; Yang, Y.-F.; Ying, G.-G. Toxic effects of Triclosan on the detoxification system and breeding of *Daphnia magna*. *Ecotoxicology* **2013**, *22*, 1384–1394. [CrossRef] [PubMed]
65. Castro, M.; Sobek, A.; Yuan, B.; Breitholtz, M. Bioaccumulation potential of CPs in aquatic organisms: Uptake and depuration in *Daphnia magna*. *Environ. Sci. Technol.* **2019**, *53*, 9533–9541. [CrossRef]
66. Dai, Z.; Xia, X.; Guo, J.; Jiang, X. Bioaccumulation and uptake routes of perfluoroalkyl acids in *Daphnia magna*. *Chemosphere* **2013**, *90*, 1589–1596. [CrossRef]
67. Baldwin, W.S.; Milam, D.L.; Leblanc, G.A. Physiological and biochemical perturbations in *Daphnia magna* following exposure to the model environmental estrogen diethylstilbestrol. *Environ. Toxicol. Chem.* **1995**, *14*, 945–952. [CrossRef]

Article

Selective Recovery of Copper from a Synthetic Metalliferous Waste Stream Using the Thiourea-Functionalized Ion Exchange Resin Puromet MTS9140

Alex L. Riley [1,2,*], Christopher P. Porter [1,3] and Mark D. Ogden [1]

[1] Separations and Nuclear Chemical Engineering Research (SNUCER) Group, Department of Chemical and Biological Engineering, University of Sheffield, Sheffield S1 3JD, UK; chris.porter@wastecare.co.uk (C.P.P.); m.d.ogden@sheffield.ac.uk (M.D.O.)
[2] Department of Geography, Geology & Environment, University of Hull, Hull HU6 7RX, UK
[3] WasteCare Ltd., Richmond House, Leeds LS25 1NB, UK
* Correspondence: a.l.riley@hull.ac.uk

Abstract: The extraction of Cu from mixed-metal acidic solutions by the thiourea-functionalized resin Puromet MTS9140 was studied. Despite being originally manufactured for precious metal recovery, a high selectivity towards Cu was observed over other first-row transition metals (>90% removal), highlighting a potential for this resin in base metal recovery circuits. Resin behaviour was characterised in batch-mode under a range of pH and sulphate concentrations and as a function of flow rate in a fixed-bed setup. In each instance, a high selectivity and capacity (max. 32.04 mg/g) towards Cu was observed and was unaffected by changes in solution chemistry. The mechanism of extraction was determined by XPS to be through reduction of Cu(II) to Cu(I) rather than chelation. Elution of Cu was achieved by the use of 0.5 M–1 M $NaClO_3$. Despite effective Cu elution (82%), degradation of resin functionality was observed, and further detailed through the application of IC analysis to identify degradation by-products. This work is the first detailed study of a thiourea-functionalized resin being used to selectively target Cu from a complex multi-metal solution.

Keywords: copper; ion exchange; thiourea; X-ray photoelectron spectroscopy (XPS); reductive extraction; resource recovery

1. Introduction

Copper is a globally indispensable metal used in a wide range of industries, especially in electronics, and the demand for copper resources have driven production rates exponentially over the past century [1]. Copper is traditionally obtained through the extraction of natural ore deposits, predominantly chalcopyrite ($CuFeS_2$) and chalcocite (Cu_2S), and subsequent processing by pyro- and hydro-metallurgical processes. However, these processes are not entirely efficient and result in unrecovered copper remaining present in flotation tailings [2]; a source which is predicted to increase as high-quality ore deposits are progressively depleted [3]. Furthermore, copper ores also contain metals such as Co, Ni, and Zn, and are associated with minerals such as pyrite (FeS_2) and sphalerite (ZnS) [4]. As such, it is possible for copper process waters to contain an abundance of other metal species, highlighting the need for the selective recovery of copper away from other first-row transition metals.

In addition to mineralogical sources of copper, recent efforts have been made in recovering copper from novel sources, including industrial residues [5], legacy waste deposits [6], sewage sludge [7], and e-waste [8]. Of these sources, e-wastes (particularly printed circuit boards (PCBs)) have received significant attention for their resource recovery potential [9–12]. Specifically, spent acid etching solution and waste PCB sludge have been candidates for resource recovery [9], with the sludge containing Fe, Cu, Ca, Sn, Al, Zn, and Cr [10]. Should a hydrometallurgical approach be taken to liberate these metals from

PCB wastes, it follows that a complex waste stream containing multiple metals would be produced. As such, the need for an effective method of selectively recovering copper away from other metals present in solutions is further emphasised.

As a complexing reagent, thiourea ($SC(NH_2)_2$) is most commonly used during the hydrometallurgical leaching of gold from ores as a less-toxic alternative to cyanide [13,14]. Further uses include the extraction of gold and silver from PCBs [15], and, more recently, it has been explored as part of a dual-lixiviant treatment process for waste activated carbon [16]. Given its high affinity for metals as a ligand, thiourea has been commercially incorporated into solid-phase extraction media, particularly ion exchange resins, though to date it has only been applied for precious metals extraction [17]. This paper describes the results of a previously unexpected interaction exhibited by the thiourea-functionalized resin Puromet MTS9140. Initially performed as part of a mixed-metal screening experiment to assess resin metal extraction behaviour, a high selectivity towards copper was observed and further investigated, the results of which are presented here in detail for the first time.

2. Materials and Methods

2.1. Solution Preparation

A mixed metal stock solution was prepared using the sulphate salts of Al(III), Co(II), Cu(II), Fe(III), Mn(II), Ni(II), and Zn(II). All metal salts used were of analytical grade and purchased from Sigma-Aldrich. Salts were dissolved in deionized water and acidified to pH 1 using H_2SO_4 such that the final concentration of each metal was 2000 mg/L. While concentrations in real waste leachates would be heterogeneous, the use of equal concentrations in this work ensured that observed differences in metal extraction were the result of resin behaviour. The resulting products of this "stock solution" were taken and diluted to produce the pregnant liquor solutions (PLS) that were used in experimental procedures (typically 200 mg/L). The PLS were adjusted to the appropriate pH using concentrated H_2SO_4 and NaOH to minimize changes in volume. Working from a stock solution ensured continuity in metal concentrations between batches of PLS, hence minimising variation in solutions between resin contacts. For the pH screening study, addition of H_2SO_4 allowed a range of acidities to be explored, from 0.01–3 M H^+. Where the effects of increased sulphate concentration were being tested, this was achieved by addition of $(NH_4)_2SO_4$, ranging from 0.02–4 M SO_4^{2-}. Eluent solutions were prepared by dissolution of $NaClO_3$ and adjustment to pH 2 using 37% HCl.

2.2. Resin Preconditioning

Puromet MTS9140 was supplied by Purolite International Ltd. and was preconditioned through contact with excess 1 M H_2SO_4 (S:L 1:50) for 24 h while being agitated on an orbital shaker. Five washing cycles using excess deionized water (S:L 1:50) ensured the removal of residual acid from the preconditioning process, and the resin was stored under deionized water until required. Samples of resin were dried at 50 °C to determine its density, which is provided in Table 1 alongside the manufacturer specifications.

Table 1. Manufacturer specifications of Puromet MTS9140 from Purolite International Ltd. (PS-DVB = Polystyrene cross-linked with Divinylbenzene).

Functional Group	Capacity (eq/L)	Polymer Matrix	Moisture (%)	Particle Size (μm)	Density (g/mL)
Thiourea	1	PS-DVB [1]	50–56	300–1200	0.308

[1] PS-DVB = Polystyrene crosslinked with Divinylbenzene.

2.3. Static Equilibrium Experiments

Static (batch) experiments were performed by contacting a fixed volume of resin with a constant volume of solution and allowing the system to equilibrate. Solutions were generated such that the effect of a range of acidities and sulphate concentrations on metal extraction could be determined.

Resin was measured out volumetrically by pipetting the resin/water slurry into a measuring cylinder, inverting the cylinder to promote particle size mixing, and allowing the resin to settle under gravity (herein referred to as 'wet settled resin'). 2 mL of wet settled resin was drained and contacted with 50 mL of PLS. Containers were placed on an orbital shaker and contacted for 24 h to equilibrate, after which the supernatant pH was measured using a calibrated silver/silver reference electrode and sampled for elemental analysis. This was achieved through dilution using a 1% HNO_3 solution prior to analysis via inductively coupled plasma optical emission spectroscopy (ICP-OES; Perkin Elmer Optima 5300 DV or Spectro Arcos model) or flame atomic absorption spectroscopy (AAS; Perkin Elmer AAnalyst 400 model). For all instruments, regular check-standards were analysed to ensure data accuracy, and instruments were recalibrated if readings were beyond 2.5% of the expected standard concentrations.

2.4. Fixed-Bed (Dynamic) Experiments

For dynamic breakthrough experiments, small-scale columns were completely packed with resin and capped at both ends with Teflon frits, resulting in a total bed volume (BV) of 5 mL wet settled resin. The columns were agitated during packing to promote homogeneous distribution of resin particle size throughout the bed. To ensure efficient mass transfer between solution and resin and to reduce the risk of 'channelling' [18] a reverse-flow setup was employed, whereby the PLS was introduced at the bottom of the vertical column. For elution studies, a smaller BV of 1.4 mL of wet settled resin was used to minimize the concentration of eluent peaks and required dilution for analysis. PLS were pumped using either a 'Heidolph Hei-Flow Value 01' pump with 'SP Quick' pump head, or a 'BioRad Econo Gradient Pump'. Verification of solution flow rates was achieved by pumping deionised water through each packed column for a set time and using the mass of water collected to calculate volumetric flow in bed volumes per hour (BV/h). Effluent solutions were collected using a 'BioRad Model 2110' fraction collector set to advance at specified time intervals and diluted using either 1% HNO_3 for ICP-OES or AAS analysis, or deionized water for ion chromatography (IC) analysis. IC analysis was performed using a Metrohm '833 Basic IC Plus' fitted with a 'Metrosep A Supp 5' column (PVA-quaternary ammonium), a carbonate eluent (4.5 mM Na_2CO_3, 803 µM $NaHCO_3$), and a 0.1 M H_2SO_4 regenerant.

2.5. Solid-State Analysis

The elemental composition of Puromet MTS9140 and the oxidation state of adsorbed Cu was determined via X-ray photoelectron spectroscopy (XPS) using a 'Kratos AXIS Supra' instrument and monochromated Al source. A small sample of Cu-loaded resin was homogenised in a clean pestle and mortar using a small amount of deionized water to form a paste. This was gently dried overnight at 50 °C to produce a fine powder of ground resin and was submitted to the Sheffield Surface Analysis Centre, where a subsample was pressed into indium foil prior to analysis. Survey scans were carried out between 1200–0 eV energy resolution and one 300 s sweep. High resolution C 1s, Cu 2p, and Cu LMM scans were collected at their appropriate energy ranges at 0.1 eV energy resolution, with a 300 s sweep for C and three 300 s sweeps for the Cu 2p and Cu LMM scans.

2.6. Breakthrough Modelling

Ion breakthrough was analysed using multiple breakthrough models commonly applied to ion exchange data; the modified dose–response (MDR), Bohart–Adams, Thomas, and Yoon–Nelson models. It is important to note is that the models were not necessarily intended to describe a solid-liquid ion exchange extraction process at the time they were developed, and as such the calculated values may not accurately describe experimental reality [19]. However, given the widespread use of such models in the field, this remains the only way to consistently compare new experimental data with existing literature, and so this approach is justified. Model fitting was performed for individual metal breakthrough in OriginPro 2020b software using non-linear regression analysis.

The MDR model is given in Equation (1) [20], where F_t is the cumulative flow-through (mL) at a given time, and a and b are model constants. From evaluating the MDR model, the maximum column loading capacity for each metal (Q_o) can also be derived using Equation (2), where m is the mass of resin used (g).

Equation (1). Modified dose response model.

$$\frac{C_t}{C_o} = 1 - \frac{1}{1 + \left(\frac{F_t}{b}\right)^a} \quad (1)$$

Equation (2). Calculation of Q_o from MDR constant b.

$$Q_o = \frac{b\, C_o}{m} \quad (2)$$

The Bohart–Adams model is given in Equation (3), where it is assumed that the rate of adsorption is dependent on both the concentration of sorbing species in solution and on the remaining capacity of the adsorbent. While originally developed for describing a gas-charcoal adsorption system [21,22] the model can also be applied to a solid phase extraction system from a solution phase. In Equation (3), K_a is the Bohart–Adams adsorption rate constant (L mg^{-1} min^{-1}), W is the column adsorption capacity (mg/g), and F represents the volumetric flow rate (L/hour).

Equation (3). Bohart–Adams model.

$$\frac{C_t}{C_o} = \left(\frac{e^{K_a C_o t}}{e^{K_a C_o t} + e^{K_a\left(\frac{W}{F}\right)} - 1}\right) \quad (3)$$

The Thomas model (Equation (4), where K_t is the model constant (L min^{-1} mg^{-1}) is also commonly applied to ion exchange breakthrough data and was originally developed to describe adsorption to a zeolite bed [23]. A high goodness-of-fit to this model would suggest that uptake is governed by mass transfer at the resin-solution interface [24].

Equation (4). Thomas model.

$$\frac{C_t}{C_o} = \frac{1}{1 + e^{\left(\frac{K_t Q_o m}{F} - K_t C_o t\right)}} \quad (4)$$

The final model fitted to breakthrough data is the Yoon–Nelson model, originally developed for describing the adsorption of gases to solid adsorbents, is presented in Equation (5) where K_{yn} is the model constant (min^{-1}), and t_{50} is the predicted time for 50% breakthrough to be–a useful parameter for understanding column operating times.

Equation (5). Yoon–Nelson model.

$$\frac{C_t}{C_o} = \frac{1}{1 + e^{K_{yn}(t_{50} - t)}} \quad (5)$$

3. Results and Discussion

3.1. Static (Batch) Extraction

When contacted with the PLS under batch conditions, Puromet MTS9140 exhibited great selectivity towards Cu across all studied pH conditions (Figure 1), with extraction remaining above 92% irrespective of increased proton concentration. The extraction of other metals present in the PLS was very low, with almost no change in extraction percentage with respect to equilibrium pH.

Increased sulphate concentration in the PLS had no observable effect on the extraction of any metal species by S914, with the high capacity for Cu removal and low co-removal of other metals reported under all sulphate concentration conditions (Figure 2), indicating a high resilience to increased ionic strength. Given that this resin was not developed by the manufacturer for Cu separations, and that true selectivity towards a single ion would be highly advantageous for treatment of mixed metal waste streams, further exploration into the surface binding of Cu was explored for this resin by using X-ray Photoelectron Spectroscopy (XPS) analysis to better understand the extraction process.

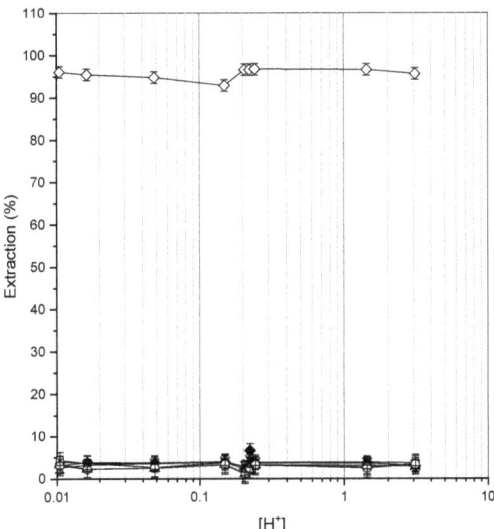

Figure 1. Extraction percentage of metal ions as a function of acid concentration on MTS9140. Al(III) = +, Co(II) = □, Cu(II) = ◇, Fe(III) = ◆, Ni(II) = ▲, Mn(II) = ×, Zn(II) = ○.

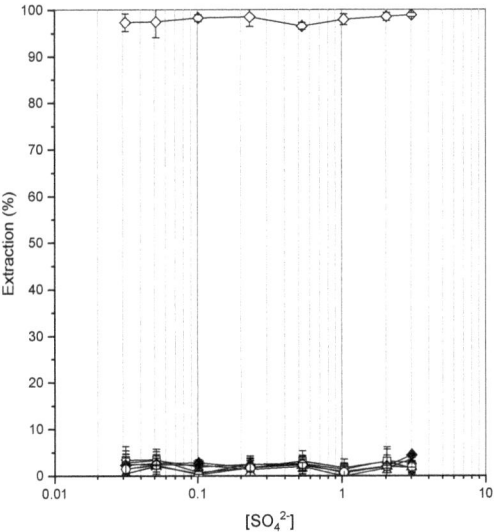

Figure 2. Extraction percentage of metal ions as a function of ammonium sulphate concentration on MTS9140 at 0.027 M H^+. Al(III) = +, Co(II) = □, Cu(II) = ◇, Fe(III) = ◆, Ni(II) = ▲, Mn(II) = ×, Zn(II) = ○.

The XPS survey scan was used to determine surface elemental composition of powdered sample of Cu-loaded S914, given as atomic % (At%) in Figure 3. As can be expected for a resin bead with a polystyrene-DVB backbone, the most abundant element quantified was C (63.77 At%). In decreasing abundance, the remaining composition was determined to be O (16.98 At%), N (7.62 At%), Si (7.10 At%), S (4.28 At%), and Cu (0.25 At%) (Figure 3). All elemental peaks were clearly detected, as evidenced by the low 'full width at half maximum' (FWHM) values reported in the table within Figure 3.

The determined binding energy of N (400 eV) is consistent with nitrogen within an organic matrix, i.e., the thiourea functional group within the resin matrix. While the presence of N was expected given the functionality of the resin, the high Si and O concentrations were not expected. The binding energies of Si $2p$ and O $1s$ peaks in Figure 3 were consistent with the presence of silica (SiO_2), yet the percentage composition of oxygen indicates that silica was not the only oxygen source present in the sample. Close examination of the S $2s$ peak revealed two components; a peak consistent with thiourea (75% of S detected), and a smaller peak (25% of S detected) attributed to the presence of sulphate–presumably residual sulphate from the PLS during loading.

Considering that the resin was loaded with excess Cu in solution, the Cu surface concentration was relatively small (0.25 At%). This was likely due to a diluting effect of the bulk resin matrix (PS-DVB), which could overshadow the presence of functional groups. The low concentration of thiourea relative to the bulk matrix meant that the photoelectrons emitted from Cu atoms (associated with thiourea groups) originated too deep within the sample to escape without further collision and were less-well detected.

In addition to the full survey scan to determine bulk elemental composition, a more detailed scan of the Cu $2p$ environment was performed to investigate the speciation of Cu when loaded to S914. Figure 4 indicates a clear symmetrical peak at 932.82 eV, peaking at approximately 214 counts per second (CPS). A second, less well-defined peak was also observed in the region of 952.5 eV, with a magnitude of 190 CPS. While the emission baseline appeared noisy during the Cu $2p$ scan, this is a direct result of the relatively low At% of Cu in the Cu-loaded resin sample (Figure 3).

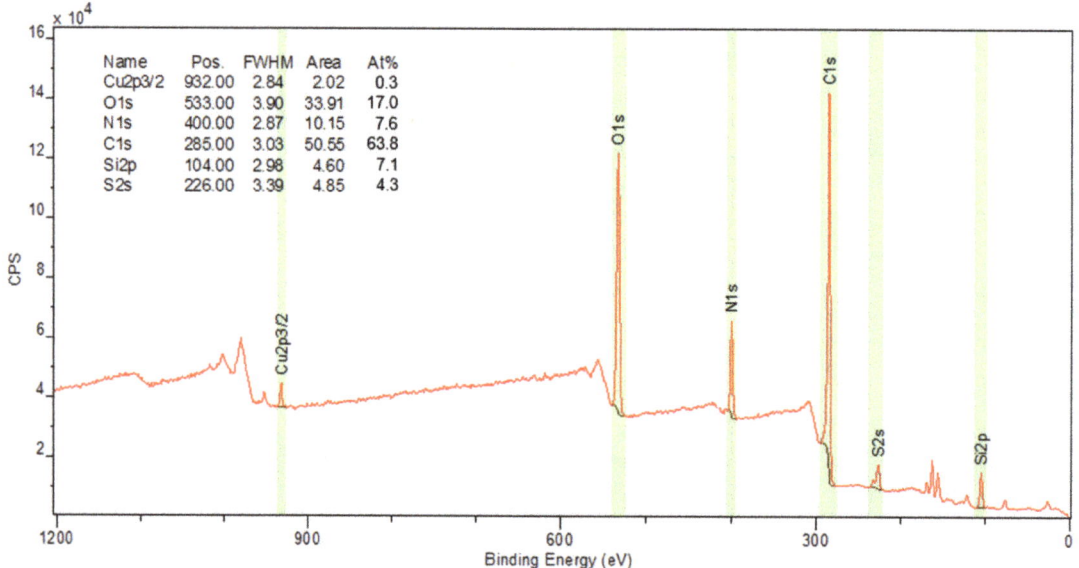

Figure 3. XPS survey scan for determination of elemental composition of Cu-loaded MTS9140 (CPS = counts per second).

The lack of Cu satellite peaks in the 940–945 eV region eliminated the possibility for Cu(II) being present within the sample [25], instead the binding energy of the Cu 2p peak was consistent with either Cu metal or Cu(I) (933 eV for both species), and the secondary peak at 952.5 eV coupled with the absence of satellite peaks provides further evidence for this [26].

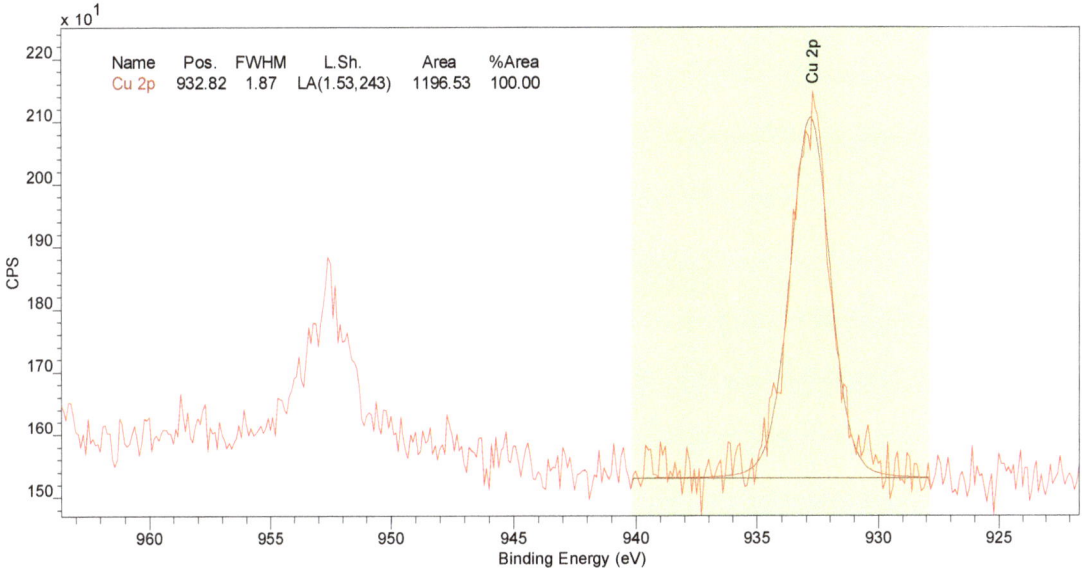

Figure 4. XPS Cu 2p scan for Cu-loaded MTS9140 (CPS = counts per second).

For Cu to be present on the resin surface as either Cu(I) or Cu metal, it follows that the Cu(II) present in solution must undergo reduction, with the functional group of the resin in turn being oxidised. Such redox behaviour between Cu(II) and solutions of thiourea has been previously observed [27], whereby Cu(II) is immediately bonded with thiourea which is in turn quickly oxidised by Cu(II) ions. The resulting Cu(I) ion produced as a result of this redox reaction is then complexed by thiourea to form a stable Cu(I)-thiourea complex, the form of which may involve one, two, or three thiourea ligands [28]. Given the reported interactions of Cu(II) ions with thiourea ligands in solutions, and given that Cu metal would be expected to give a sharper and more asymmetric peak than observed in Figure 4 [29], it is proposed that S914 removes Cu from solution through reduction to cuprous Cu(I).

3.2. Fixed-Bed Adsorption

Under dynamic operation, S914 continued to exhibit exclusive Cu selectivity and extraction from the PLS as evidenced by the almost immediate breakthrough of all other ions from the column, which reached complete breakthrough within the first five BV throughput (Figure 5). Numerical modelling for these metals indicated very low loading capacities for these ions (2.02–2.15 mg/g, MDR, Table 2), but considering the speed at which these metals broke through and considering the lack of displacement following complete breakthrough, this is likely an overestimation of loading capacity.

Cu breakthrough began to occur at around 5 BV throughput and gradually increased, following a slightly sigmoidal pattern, until reaching a concentration ration of 0.92 at BV 80 (Figure 5). While the breakthrough curve was not sharp, Cu extraction appeared unaffected by other metal ions in solution and was best-defined by the MDR model, which indicated a loading capacity of 19.84 mg/g.

Changing solution flow rate to 5 BV/h had little effect on the uptake of Co, Fe, Mn, Ni, and Zn, which passed through the column with little-to-no interaction with the thiourea-functionalised resin (Figure 6). For Cu, which was exclusively extracted by S914, the breakthrough profile appeared sharper than it did when loaded at 10 BV/h, which was confirmed by an associated increase in the MDR model constant 'a' (Table 3). Complete Cu breakthrough was also achieved earlier at the 5 BV/h flow rate, where Cu concentration ratio reached 1 at 68 BV throughput and corresponded to a Q_o value of 20.68–22.77 mg/g according to the breakthrough modelling (MDR and Thomas model had equal goodness-of-fit, R^2 = 0.998).

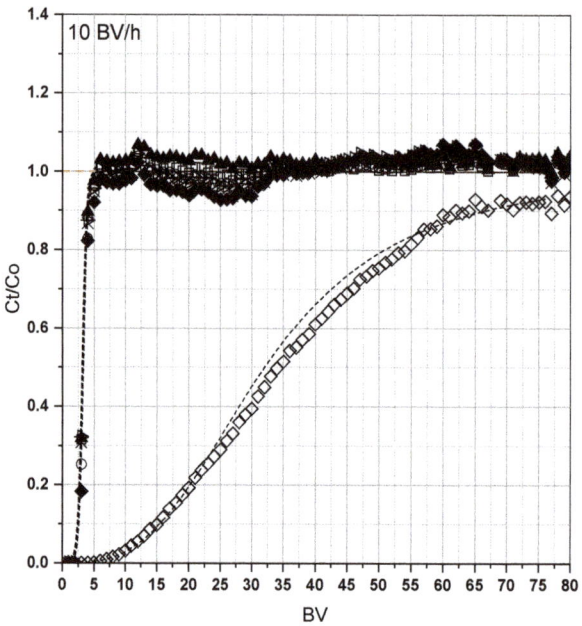

Figure 5. Breakthrough curves of metals from PLS pumped through MTS9140 at 10 BV/h (pH 1.56). Al = ▷, Co = □, Cu = ◇, Fe = ◆, Mn = ×, Ni = ▲, Zn = ○. Dotted lines = best-fitting breakthrough model (see Table 2).

Table 2. Breakthrough model parameters for MTS9140 at 10 BV/hour flow rate.

	Modified Dose Response				Bohart–Adams			Thomas			Yoon–Nelson		
	a	b	Q_o	R^2	K_a	W	R^2	K_t	Q_o	R^2	K_{yn}	t_{50}	R^2
Al	9.15	15.45	2.10	0.996	0.14	3.27	0.996	0.14	2.12	0.996	0.49	18.22	0.998
Co	9.64	15.44	2.15	0.999	0.14	3.34	0.999	0.14	2.17	0.999	0.52	18.20	0.996
Cu	3.01	160.26	19.84	0.998	0.005	33.07	0.989	0.005	21.20	0.989	0.02	199.98	0.989
Fe	9.97	16.48	2.13	0.980	0.16	3.31	0.979	0.16	2.15	0.979	0.53	19.44	0.995
Mn	9.06	15.57	2.02	0.986	0.15	3.14	0.985	0.15	2.04	0.985	0.49	18.36	0.997
Ni	10.07	15.37	2.10	0.999	0.15	3.26	0.999	0.15	2.11	0.999	0.54	18.11	0.997
Zn	10.07	15.37	2.01	0.999	0.14	3.27	0.997	0.14	2.12	0.997	0.48	18.96	0.999

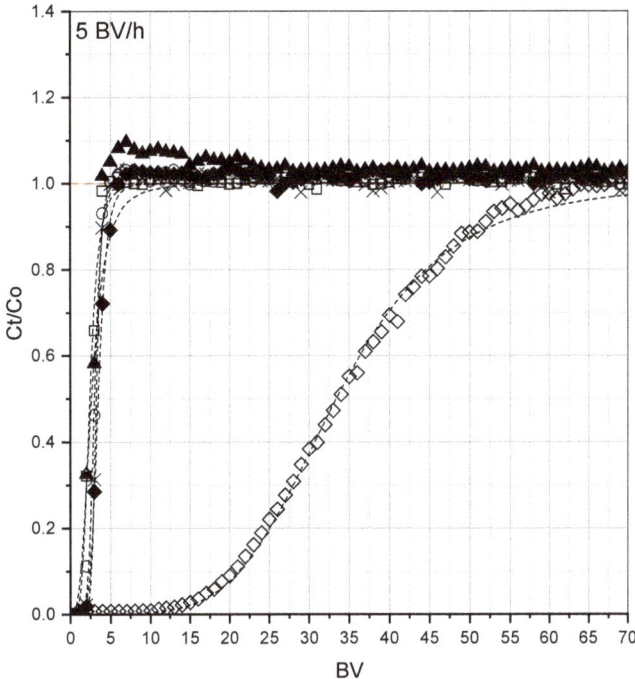

Figure 6. Breakthrough curves of metals from PLS pumped through MTS9140 at 5 BV/h (pH 1.56). Co = □, Cu = ◇, Fe = ◆, Mn = ×, Ni = ▲, Zn = ○. Dotted lines = best-fitting breakthrough model (see Table 3).

Table 3. Breakthrough model parameters for MTS9140 at 5 BV/hour flow rate.

	Modified Dose Response				Bohart-Adams			Thomas			Yoon-Nelson		
	a	b	Q_o	R^2	K_a	W	R^2	K_t	Q_o	R^2	K_{yn}	t_{50}	R^2
Co	4.49	12.15	1.69	0.994	0.05	2.72	0.995	0.05	1.76	0.995	0.18	26.52	0.992
Cu	4.81	167.02	20.68	0.998	0.004	35.13	0.998	0.004	22.77	0.998	0.01	360.37	0.998
Fe	6.45	17.32	2.23	0.999	0.05	3.50	0.997	0.05	2.27	0.997	0.18	36.91	0.996
Mn	10.29	16.18	2.10	0.999	0.09	3.26	0.999	0.09	2.12	0.999	0.28	34.22	0.998
Ni	3.09	13.20	1.80	0.987	0.04	2.85	0.956	0.04	1.84	0.956	0.17	27.37	0.992
Zn	7.35	15.07	1.87	0.995	0.07	3.05	0.999	0.07	1.98	0.999	0.23	31.79	0.998

When reduced to 2 BV/h, a further sharpening of the breakthrough of ions was observed. For all metals other than Cu, almost immediate complete breakthrough occurred, after which concentration ratios plateaued at values between 1 and 1.1 (Figure 7). It is important to note that a decision was made to increase sampling resolution during the 2 BV/h flow rate study to increase the number of data points during the frontal portion of the breakthrough curve. Given the limited space for sample collection in fraction collectors, this resulted in only the first 40 BV being sampled. Nevertheless, sufficient Cu breakthrough was captured to allow effective modelling and comparison to other flow rates. Cu was first detected in effluent solutions beyond 25 BV throughput (Figure 7), much later than observed for 10 and 5 BV/h (BV 5 and 10, respectively).

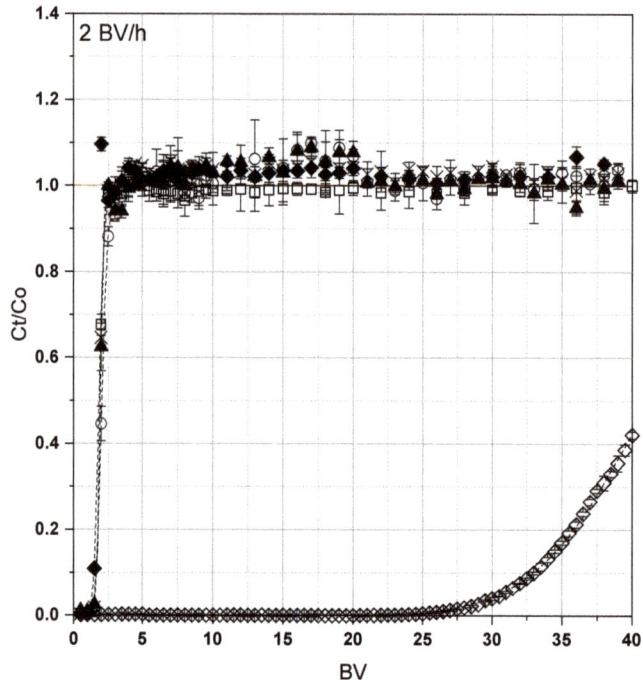

Figure 7. Breakthrough curves of metals from PLS pumped through MTS9140 at 2 BV/h (pH 1.56). Co = ☐, Cu = ◇, Fe = ◆, Mn = ×, Ni = ▲, Zn = ○. Dotted lines = best-fitting breakthrough model (see Table 4).

Table 4. Breakthrough model parameters for MTS9140 at 2 BV/hour flow rate.

	Modified Dose Response				Bohart-Adams			Thomas			Yoon-Nelson		
	a	b	Q_o	R^2	K_a	W	R^2	K_t	Q_o	R^2	K_{yn}	t_{50}	R^2
Co	16.67	9.56	1.46	0.996	0.09	2.25	0.997	0.09	1.46	0.997	0.333	57.52	0.999
Cu	9.74	206.45	32.04	0.999	0.002	48.75	0.997	0.002	31.66	0.997	0.01	1223.9	0.997
Fe	10.46	9.16	0.94	0.999	0.07	1.48	0.999	0.07	0.96	0.999	NA	NA	NA
Mn	16.14	9.63	1.08	0.999	0.10	1.66	0.999	0.10	1.08	0.999	0.30	57.72	0.999
Ni	16.57	9.69	1.18	0.999	0.09	1.81	0.999	0.09	1.18	0.999	0.29	57.94	0.999
Zn	9.97	10.25	1.13	0.996	0.06	1.73	0.996	0.06	1.13	0.996	0.17	61.52	0.996

As was the case for batch pH and sulphate concentration screening, no published data exists studying the fixed-bed adsorption of metals using Puromet MTS9140. Under dynamic operation, MTS9140 displayed the same Cu-selective extraction properties as under previous batch experiments. All other metals present in the PLS reached complete breakthrough within the first 5 BV and exhibited very little overshoot ($C_t/C_o > 1$). The brief period of overshoot observed in Figure 6, which increased slightly as flow rate decreased (Figure 7), indicating that, upon entry to the column, metals were momentarily adsorbed but rapidly displaced as Cu began to load. The fact that all other metals reach complete breakthrough long before Cu indicated that it was not competition for remaining active sites that led to the displacement of metals, but instead that Cu was able to replace metal counter ions immediately upon contact and before resin saturation.

All breakthrough models were able to describe metal loading to MTS9140 with a high degree of accuracy (Tables 2–4), with the MDR model generally outperforming other models for Cu. Results of MDR modelling revealed Cu operating capacities of 19.84 mg/g at 10 BV/h, 20.68 mg/g at 5 BV/h, and 32.04 mg/g at 2 BV/h. While Cu capacity at the high and intermediate flow rates were comparable, a notable increase in capacity was observed at the lowest flow rate. This was likely the result of a more effective extraction at the low flow rate, evidenced by delayed breakthrough point and longer column half-lives calculated by the Yoon–Nelson model (20.4 min) when compared to 5 BV/h (6 min) and 10 BV/h (3.3 min) operation.

3.3. Resin Elution

The proposed reductive extraction mechanism of Puromet MTS9140 suggested an oxidative eluent to be applied to liberate Cu(I) as Cu(II) from the resin. While no literature exists on the elution of Cu from Puromet MTS9140 specifically, a limited number of articles do exist that explore the elution of Cu from non-commercial dual-functionalized resins containing, among other groups, thiourea functionality. One such paper reports effective batch Cu recovery from a thiourea/acyl bifunctional resin using concentrated nitric acid (equivalent to 3.0 M) as an eluent [30], and was initially explored (see Supplementary Material).

Based on initial experimentation, it was deemed that the use of concentrated nitric acid as an eluent for MTS9140 would be unsuitable for two main factors; (1) the observed formation of nitrous gases could hinder process scale-up, and (2) the safety concerns of handling large volumes of highly concentrated nitric acid after process scale-up. However, given the presence of Cu(I) on the thiourea resin surface, an oxidative elution approach remained a possible avenue of exploration, and so sodium chlorate ($NaClO_3$) was chosen for further elution studies, given the higher reduction potential of chlorate (1.152–1.451 V) over nitrate (0.803–0.934 V) in acidic media [31]. In addition to acting as a stronger oxidising agent, the use of $NaClO_3$ over HNO_3 allowed elution to be performed under less-acidic conditions.

The elution profile of Cu from S914 using a 0.5 M solution of $NaClO_3$ at pH 2 is presented in Figure 8. The concentration of Cu in effluent solutions increased sharply beyond 4 BV throughput, reaching a maximum concentration of 511 mg/L at 20 mL (~14 BV) throughput. Following peak maximum, Cu concentration exhibited a steep decline, which gradually levelled out to below 10 mg/L by the end of the experimental run; allowing the complete profile to be captured. Integration of the area below the curve (Table 5) revealed effective Cu elution by this eluent, with an overall Cu recovery percentage of 78.91%.

Table 5. Details of Cu elution investigations using $NaClO_3$ (FWHM provided for comparison of peak widths).

[$NaClO_3$]	Cu Loaded (mg/mL)	Bed Volume (mL)	Total Cu on Bed (mg)	Cu Recovered (mg)	FWHM (mL)	Recovery Efficiency (%)
0.5 M	8.64	1.4	12.10	9.55	16.2	78.91
1.0 M	8.36	1.4	11.70	9.58	13.2	81.86

Doubling the concentration of sodium chlorate had the effect of increasing the maximum Cu concentration to 612 mg/L (Figure 9). Despite the higher Cu concentration during peak maximum, the width of the elution peak (FWHM) was smaller than that in Figure 8, resulting in a recovery efficiency of 81.86%; fairly similar to the Cu recovery using 0.5 M $NaClO_3$ (Table 5). Following elution using 1 M $NaClO_3$, a slight but notable colour change was observed, with the 1M-contacted S914 taking on a grey hue when compared to the 0.5 M-contacted resin. It is theorised that this grey colouration was the result of copper oxide formation on the surface of the resin bead.

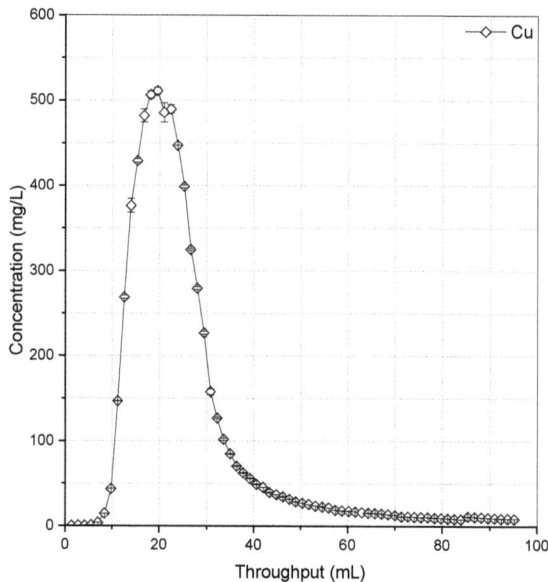

Figure 8. Elution of Cu from MTS9140 using 0.5 M NaClO3 at pH 2 (HCl matrix, 2 BV/h).

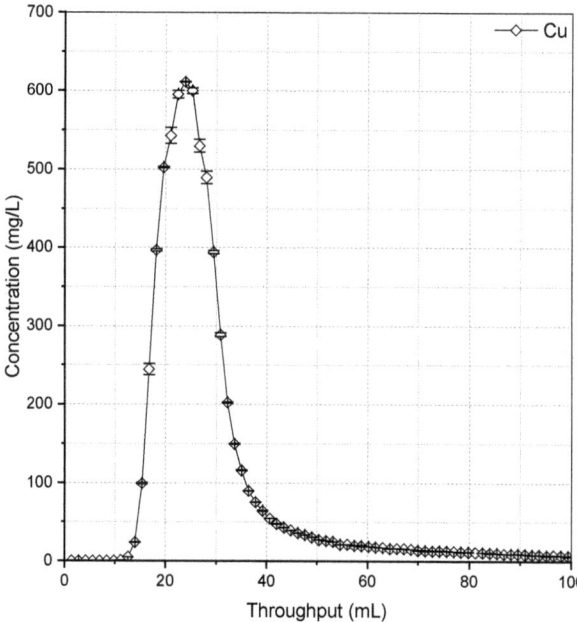

Figure 9. Elution of Cu from MTS9140 using 1 M NaClO$_3$ at pH 2 (HCl matrix, 2 BV/h).

No gas formation was observed during chlorate elution, which was also evidenced by the more regular shape of elution profiles when compared to those obtained during HNO$_3$ elution (see Supplementary Information). Cu elution profiles were fairly symmetrical with a steep frontal curve and slightly longer tail-end. This curve profile is often observed where effective elution is achieved [32–34], and implies a favourable exchange (or removal) of the target ion in favour of the eluent counter-ion.

3.4. Resin Reusability

To determine the reusability of MTS9140 for the selective extraction of Cu, repeated loading, elution, and preconditioning cycles were performed. The 0.5 M NaClO$_3$ eluent was used given its effectiveness for Cu recovery. While the 1 M eluent did have a marginally higher recovery efficiency, the lack of resin discoloration indicated that 0.5 M NaClO$_3$ was a better option for investigating reusability. The resulting breakthrough curves are presented in Figure 10.

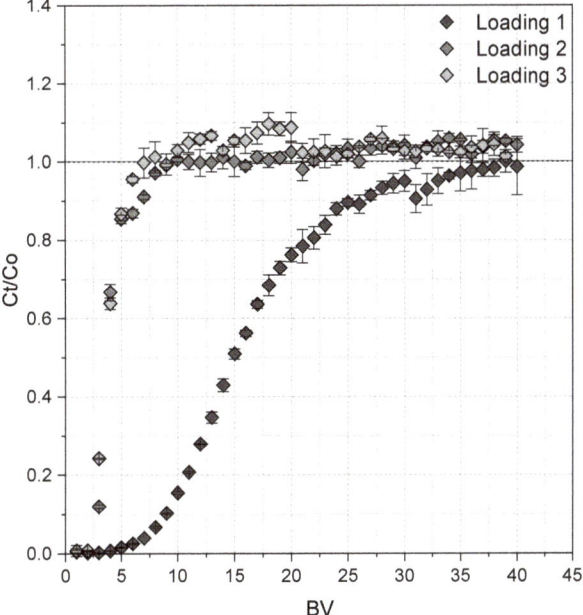

Figure 10. Breakthrough curves of Cu from MTS9140 over repeated loading cycles (1.4 mL BV, 5 BV/h; 400 mg/L Cu, pH 1.55).

During the first loading cycle, typical Cu adsorption behaviour was observed, with complete breakthrough encountered at 40 BV throughput (Figure 10). Integration of the area above the breakthrough curve (with an upper boundary of influent PLS concentration) indicated the adsorption of 7.95 mg Cu during the first loading cycle. Following Cu elution, the resin bed was loaded again using the 400 mg/L Cu solution, but exhibited very low adsorption of Cu, with only 1.55 mg of Cu removed before complete breakthrough occurred during both the second and third loading cycles (Table 6).

Comparison of elution curves using 0.5 M NaClO$_3$ (Figure 11) revealed that for the first cycle, where 7.95 mg Cu was loaded to the resin bed, 6.79 mg was recovered, equating to a recovery efficiency of 85.34% (Table 6). The elution profile generated increased sharply, reaching a maximum Cu concentration of 418 mg/L after 0.028 L, before decreasing in an almost-symmetrical fashion, indicating effective desorption.

The considerably lower extent of Cu extraction during the second and third loading cycles (Figure 10) was reflected in the elution profiles for the respective cycles (Figure 11). However, for elution cycle two, the recovery efficiency was consistent with recovery during cycle one (88.46 and 85.34, respectively; Table 6). While the extent of Cu loading between cycle one and two was similar, the mass of Cu eluted differed greatly, with a decline in recovery efficiency from 88.46% to only 46.27% during elution cycle three (Table 6).

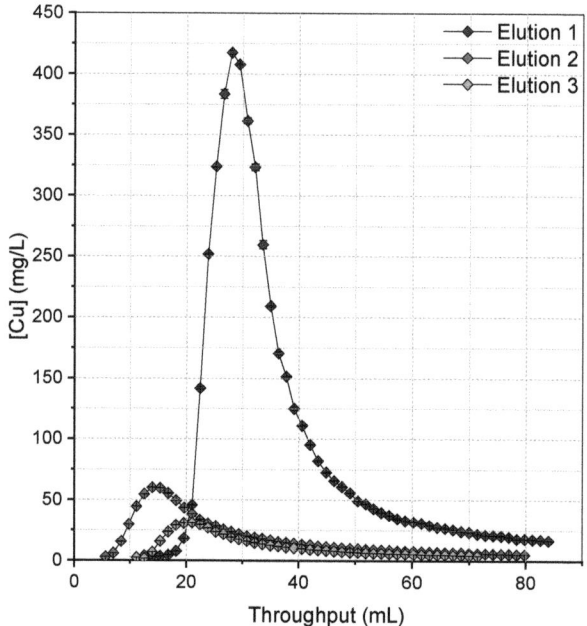

Figure 11. Elution of Cu from MTS9140 using 0.5 M NaClO$_3$ (pH 2, HCl matrix, 2 BV/h) over repeated elution cycles.

Table 6. Masses of Cu loaded and eluted from MTS9140 over multiple operational cycles.

Cycle	Loading (mg)	Elution (mg)	Recovery (%)
1	7.95	6.79	85.34
2	1.55	1.37	88.46
3	1.55	0.72	46.27

The lower mass of Cu recovered from S914 during the third elution cycle could potentially be explained by the progressive loss of resin functionality during elution cycles. It could be assumed that as the eluent contacts the exterior of resin beads before diffusing through the resin pores, the functional groups at the surface of each bead would be the first to be degraded, explaining the loss of capacity between cycle one and two. Assuming this 'shrinking-core' hypothesis for loss of functionality is true, it holds that during the later elution cycles, the remaining functionality is located deeper within the resin beads and therefore less accessible to the eluent. While the ionic radii of Cu(I) (0.46–0.77 Å) is substantially smaller than that of ClO$_3^-$ (1.71 Å), it is unlikely that Cu residing in pores inaccessible to ClO$_3-$ was responsible for this observation given the flexibility of gelled polymers [18]. Instead, this is more likely a result of kinetic limitations inherent to column operation, i.e., the limited residence time within the column is hindering sufficient mass transfer between the bulk eluent and resin, and in doing so reduces eluent efficiency [35]. It is expected that such effects are amplified when coupled with functionality degradation on resin outer surfaces.

3.5. Functionality Degradation

It is evident that while Cu can successfully and efficiently be recovered from MTS9140, a cuprous oxidation approach to elution is unsuitable for maintaining the functionality of the resin for reuse. To better understand this degradation of MTS9140, Cu elution using 0.5 M $NaClO_3$ was repeated on a Cu-loaded column, with effluent bed volumes being sampled and analysed by IC.

During elution of Cu, an increase in pH was observed in effluent solutions, peaking at pH 3.13 at 7 BV throughput (Figure 12); an increase of 1.18 pH units from the native pH of the eluent used (pH 1.95). A lag of 5 BV was observed until peak Cu elution, which occurred after 12 BV and reached a concentration of 604 mg/L.

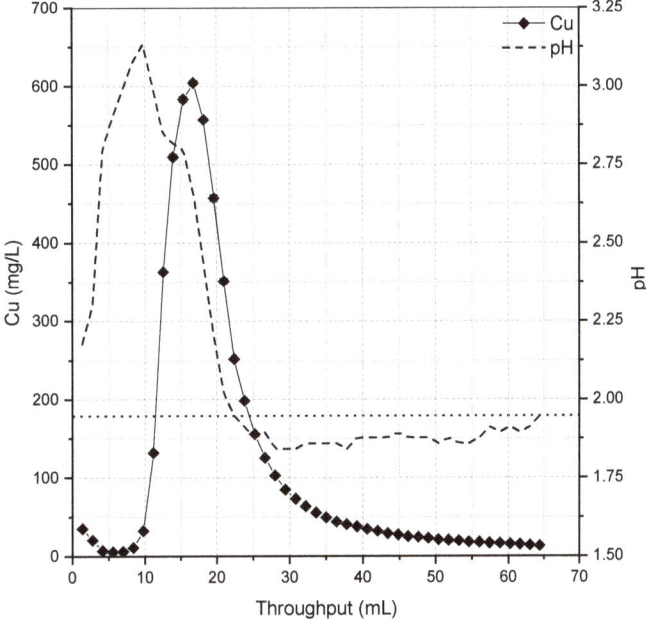

Figure 12. Cu concentration and pH of effluent solutions during elution of Cu from MTS9140 at 2 BV/h using 0.5 M $NaClO_3$ (pH 1.95, HCl media, dotted horizontal line represents pH of eluent).

Given that the chlorate ion is fundamental to the oxidation of the Cu(I) centre, it was theorised that a peak in chloride would occur alongside the Cu elution peak, and so this was analysed for by IC, as well as sulphate concentrations. The concentrations of Cl^- and SO_4^{2-} are presented alongside the Cu elution profile in Figure 13. A peak in chloride concentration was observed simultaneously with peak Cu elution, with a maximum concentration of 0.02 M Cl^- (714 mg/L). A substantial peak in sulphate concentration (maximum 0.085 M SO_4^{2-} (5646 mg/L)) was also detected, occurring prior to the peak in Cu and Cl, and before the period of increased pH.

The peak in Cl^- is not unexpected, and the parallel occurrence of Cu and Cl^- elution confirmed that the method of Cu liberation is a redox interaction between ClO_3^- and Cu(I), whereby ClO_3^- is reduced to Cl^- and Cu(I) oxidised to Cu(II). However, the large spike in sulphate concentration is particularly significant given that no sulphate was introduced to the system during the elution cycle. While the PLS used for loading did contain sulphate, a thorough rinse cycle using 18 MΩ deionised water was performed prior to elution such that residual sulphate concentration was below 0.002 M prior to the start of the sulphate peak in Figure 13.

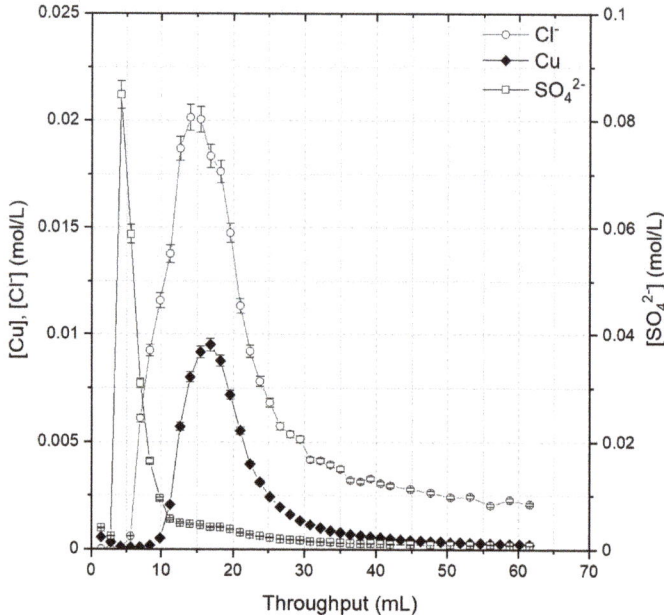

Figure 13. Concentrations of Cu, Cl$^-$, and SO$_4^{2-}$ in effluent solutions during elution of S914 at 2 BV/h using 0.5 M NaClO$_3$ (pH 1.95, HCl media, same Cu elution profile as presented in Figure 12).

It is important to note that anion concentrations presented in Figure 13 have been corrected to consider the concentrations in both the calibration blank and eluent solution, such that the presented data represents only additional anions introduced to the system. Therefore, the sulphate peak detected in effluent solutions could have only originated as a result of resin degradation, and it is likely that this degradation is responsible for the diminished capacity observed during cyclic adsorption (Table 6). While it is possible that other processes may play a part in this reduced performance, such as the possible presence of a passive CuO film during secondary loading cycles, the spike in sulphate occurring prior to Cu elution (Figure 13) indicates that functionality degradation is the key driver of this.

While not specifically recorded for chlorate, the oxidation of thiourea compounds by other halogenic oxidants such as bromate, chlorite, and iodate has been shown to form sulphate through substitution of sulphur for oxygen on the thiourea group, and subsequent sulphur oxidation [36]. It is therefore expected that oxidation by chlorate is also able to produce such by-products during degradation of thiourea functional groups.

4. Conclusions

The extractive behaviour of a thiourea-functionalized resin, Puromet MTS9140, from an acidic mixed-metal system was studied under static and dynamic conditions, where a highly selective behaviour towards Cu extraction was observed. The mechanism of extraction was elucidated to be via reduction of Cu(II) to Cu(I) by the thiourea functionality, and an efficient oxidative elution approach was used to recover a concentrated Cu product stream under ambient conditions, albeit at expense of resin reusability. This holds potential benefit in the separation of Cu from other base metals and will aid in the development of a cost-effective Cu recovery process for a variety of existing primary and secondary sources, including emerging Cu sources in waste reprocessing.

Copper-based catalysts are widely used in chemical industries, and in addition to the mining sector, this resin may prove useful in sectors such as the pharmaceutical industry in their recent move away from using precious metal catalysts in favour of more environmentally-benign copper catalysts [37–39]. Given the tight regulations of Cu content in end-products, this target-specific resin may be of particular interest for this application also.

The issue of resin reusability brings to light opportunities in further research. The application of Puromet MTS9140 for Cu recovery from real waste and/or ore leachates would be beneficial for determining industrial applications of this resin and would allow for further optimisation of experimental parameters to suit particular needs, particularly in relation to heterogeneous metal concentrations. Further to this, the exploration of alternative low-cost materials to use as the backbone in thiourea-functionalized adsorbents would be of particular interest to explore further, given the impacts of oxidative Cu recovery on extractive performance. Ongoing research into the functionalization of silica products for metal recovery in other industries (e.g., [40,41]) may offer a solution to the issue of single-use polystyrene-DVB resins; improving overall sustainability in the process through more environmentally-conscious disposal options.

Supplementary Materials: The following are available online at https://www.mdpi.com/article/10.3390/eng2040033/s1, Nitric Acid Elution of Cu from Puromet MTS9140; Figure S1. Breakthrough curve of Cu from MTS9140 (5 mL BV, 5 BV/h, 400 mg/L Cu, pH 1.55); Figure S2. Comparison of Cu elution profiles from MTS9140 using 3 M HNO_3 at 2 BV/h (D = dynamically-loaded resin, B = batch-loaded resin); Table S1. Details of Cu elution investigations using HNO_3 (FWHM provided for comparison of peak widths); Figure S3. Elution of Cu from MTS9140 using 1 M HNO_3 at 2 BV/h.

Author Contributions: A.L.R.: Conceptualization, Methodology, Formal Analysis, Investigation, Data Curation, Writing—Original Draft, Visualization. C.P.P.: Validation, Resources, Writing—Review & Editing. M.D.O.: Conceptualization, Methodology, Validation, Writing—Review & Editing, Supervision, Funding acquisition. All authors have read and agreed to the published version of the manuscript.

Funding: This work was completed as part of a Doctoral Training Partnership PhD program (A.L. Riley) co-funded by the Engineering and Physical Sciences Research Council (EPSRC) and The University of Sheffield.

Data Availability Statement: The data presented in this study are available on request from the corresponding author. The data are not publicly available at present.

Acknowledgments: The authors would like to acknowledge the members of the Separations and Nuclear Chemical Engineering Research (SNUCER) group at the University of Sheffield who provided support in understanding the results presented in this work. Additionally, Will Mayes of the University of Hull is thanked for provision of ICP-OES analysis for static screening experiments, and Deborah Hammond of the Sheffield Surface Analysis Centre is thanked for XPS analysis. Purolite Ltd. are thanked also for donation of a range of ion exchange resins used as part of wider experimentation.

Conflicts of Interest: The authors declare no conflict of interest. The funders had no role in the design of the study; in the collection, analyses, or interpretation of data; in the writing of the manuscript, or in the decision to publish the results.

References

1. Northey, S.; Haque, N.; Mudd, G. Using sustainability reporting to assess the environmental footprint of copper mining. *J. Clean. Prod.* **2013**, *40*, 118–128. [CrossRef]
2. Han, B.; Altansukh, B.; Haga, K.; Stevanović, Z.; Jonović, R.; Avramović, L.; Urosević, D.; Takasaki, Y.; Masuda, N.; Ishiyama, D.; et al. Development of copper recovery process from flotation tailings by a combined method of high-pressure leaching-solvent extraction. *J. Hazard. Mater.* **2018**, *352*, 192–203. [CrossRef]
3. Deng, S.; Gu, G.; Ji, J.; Xu, B. Bioleaching of two different genetic types of chalcopyrite and their comparative mineralogical assessment. *Anal. Bioanal. Chem.* **2018**, *410*, 1725–1733. [CrossRef]
4. Owusu, C.; Abreu, S.B.; Skinner, W.; Addai-Mensah, J.; Zanin, M. The influence of pyrite content on the flotation of chalcopyrite/pyrite mixtures. *Miner. Eng.* **2014**, *55*, 87–95. [CrossRef]
5. Gouvea, L.R.; Morais, C.A. Development of a process for the separation of zinc and copper from sulfuric liquor obtained from the leaching of an industrial residue by solvent extraction. *Miner. Eng.* **2010**, *23*, 492–497.
6. Crane, R.A.; Sapsford, D.J. Towards Greener Lixiviants in Value Recovery from Mine Wastes: Efficacy of Organic Acids for the Dissolution of Copper and Arsenic from Legacy Mine Tailings. *Minerals* **2018**, *8*, 383. [CrossRef]

7. Bezzina, J.P.; Ruder, L.R.; Dawson, R.; Ogden, M. Ion exchange removal of Cu(II), Fe(II), Pb(II) and Zn(II) from acid extracted sewage sludge—Resin screening in weak acid media. *Water Res.* **2019**, *158*, 257–267. [CrossRef]
8. Shuva, M.; Rhamdhani, M.; Brooks, G.; Masood, S.; Reuter, M. Thermodynamics data of valuable elements relevant to e-waste processing through primary and secondary copper production: A review. *J. Clean. Prod.* **2016**, *131*, 795–809. [CrossRef]
9. Min, X.; Luo, X.; Deng, F.; Shao, P.; Wu, X.; Dionysiou, D.D. Combination of multi-oxidation process and electrolysis for pretreatment of PCB industry wastewater and recovery of highly-purified copper. *Chem. Eng. J.* **2018**, *354*, 228–236. [CrossRef]
10. Huang, Z.; Xie, F.; Ma, Y. Ultrasonic recivery of copper and iron through the simultaneous utilization of Printed Circuit Boards (PCB) spent acid etching solution and PCB waste sludge. *J. Hazard. Mater.* **2011**, *185*, 155–161.
11. Hu, S.-H.; Hu, S.-C.; Fu, Y.-P. Resource Recovery of Copper-Contaminated Sludge with Jarosite Process and Selective Precipitation. *Environ. Prog. Sustain. Energy* **2012**, *31*, 379–385.
12. Arshadi, M.; Mousavi, S.; Rasoulnia, P. Enhancement of simultaneous gold and copper recovery from discarded mobile phone PCBs using Bacillus megaterium: RSM based optimization of effective factors and evaluation of their interactions. *Waste Manag.* **2016**, *57*, 158–167. [CrossRef]
13. Li, J.; Miller, J.D. A review of gold leaching in acid thiourea solutions. *Miner. Process. Extr. Met. Rev.* **2006**, *27*, 177–214. [CrossRef]
14. Yang, X.; Moats, M.S.; Miller, J.D.; Wang, X.; Shi, X.; Xu, H. Thiourea–thiocyanate leaching system for gold. *Hydrometallurgy* **2011**, *106*, 58–63. [CrossRef]
15. Jing-Ying, L.; Xiu-Li, X.; Wen-Quan, L. Thiourea leaching gold and silver from the printed circuit boards of waste mobile phones. *Waste Manag.* **2012**, *32*, 1209–1212. [CrossRef] [PubMed]
16. Adams, C.R.; Porter, C.P.; Robshaw, T.J.; Bezzina, J.P.; Shields, V.R.; Hides, A.; Bruce, R.; Ogden, M.D. An alternative to cyanide leaching of waste activated carbon ash for gold and silver recovery via synergistic dual-lixiviant treatment. *J. Ind. Eng. Chem.* **2020**, *92*, 120–130. [CrossRef]
17. Silva, R.A.; Hawboldt, K.; Zhang, Y. Application of resins with functional groups in the separation of metal ions/species—A review. *Miner. Process. Extr. Metall. Rev.* **2018**, *39*, 395–413. [CrossRef]
18. Harland, C.E. *Ion Exchange: Theory and Practice*, 2nd ed.; Royal Society of Chemistry: Cambridge, UK, 1994.
19. Amphlett, J.; Ogden, M.; Foster, R.I.; Syna, N.; Soldenhoff, K.; Sharrad, C. Polyamine functionalised ion exchange resins: Synthesis, characterisation and uranyl uptake. *Chem. Eng. J.* **2018**, *334*, 1361–1370. [CrossRef]
20. Yan, G.; Viraraghavan, T.; Chen, M. A New Model for Heavy Metal Removal in a Biosorption Column. *Adsorpt. Sci. Technol.* **2001**, *19*, 25–43. [CrossRef]
21. Bohart, G.S.; Adams, E.Q. Some aspects of the Behaviour of Charcoal with Respect to Chlorine. *J. Am. Chem. Soc.* **1920**, *42*, 523–544.
22. Hamdaoui, O. Removal of copper(II) from aqueous phase by Purolite C100-MB cation exchange resin in fixed bed columns: Modeling. *J. Hazard. Mater.* **2009**, *161*, 737–746. [CrossRef]
23. Thomas, H.C. Heterogeneous Ion Exchange in a Flowing System. *J. Am. Chem. Soc.* **1944**, *66*, 1664–1666. [CrossRef]
24. Calero, M.; Hernáinz, F.; Blázquez, G.; Tenorio, G.; Martín-Lara, M. Study of Cr (III) biosorption in a fixed-bed column. *J. Hazard. Mater.* **2009**, *171*, 886–893. [CrossRef]
25. ThermoScientific. XPS: Copper. 2020. Available online: https://xpssimplified.com/elements/copper.php (accessed on 2 February 2020).
26. Hayez, V.; Franquet, A.; Hubin, A.; Terryn, H. XPS study of the atmospheric corrosion of copper alloys of archaeological interest. *Surf. Interface Anal.* **2004**, *36*, 876–879. [CrossRef]
27. Krzewska, S.; Podsiadly, H.; Pajdowski, L. Studies on the Reaction of Copper(II) with thiourea—III: Equilibrium and Stability Constants in Cu(II)-Thiourea-HClO4 Redox System. *J. Inorg. Nucl. Chem.* **1980**, *42*, 89–94.
28. Gaur, A.; Shrivastava, B.D.; Srivastava, K.; Prasad, J.; Raghuwanshi, V.S. X-ray absorption fine structure study of multinuclear copper(I) thiourea mixed ligand complexes. *J. Chem. Phys.* **2013**, *139*, 034303. [CrossRef] [PubMed]
29. Biesinger, M.C.; Lau, L.W.; Gerson, A.R.; Smart, R.S. Resolving surface chemical states in XPS analysis of first row transition metals, oxides and hydroxides: Sc, Ti, V, Cu and Zn. *Appl. Surf. Sci.* **2010**, *257*, 887–898. [CrossRef]
30. Huang, X.; Cao, X.; Wang, W.; Zhong, H.; Cao, Z. Preparation of a novel resin with acyl and thiourea groups and its properties for Cu(II) removal from aqueous solution. *J. Environ. Manag.* **2017**, *204*, 264–271. [CrossRef]
31. Vanýsek, P. Electrochemical Series. In *CRC Handbook of Chemistry and Physics*; Haynes, W.M., Ed.; CRC Press: Boca Raton, FL, USA, 2010; pp. 20–29.
32. Korak, J.A.; Huggins, R.; Arias-Paic, M. Regeneration of pilot-scale ion exchange columns for hexavalent chromium removal. *Water Res.* **2017**, *118*, 141–151. [CrossRef]
33. Jeffrey, M.; Hewitt, D.; Dai, X.; Brunt, S. Ion exchange adsorption and elution for recovering gold thiosulphate from leach solutions. *Hydrometallurgy* **2010**, *100*, 136–143. [CrossRef]
34. Tavakoli, H.; Sepehrian, H.; Semnani, F.; Samadfam, M. Recovery of uranium from UCF liquid waste by anion exchange resin CG-400: Breakthrough curves, elution behavior and modeling studies. *Ann. Nucl. Energy* **2013**, *54*, 149–153. [CrossRef]
35. Oliveira, L.M.; Brites, L.M.; Bustamante, M.C.C.; Parpot, P.; Teixeira, J.A.; Mussatto, S.I.; Barboza, M. Fixed-Bed Column Process as a Strategy for Separation and Purification of Cephamycin C from Fermented Broth. *Ind. Eng. Chem. Res.* **2015**, *54*, 3018–3026. [CrossRef]

36. Sahu, S.; Sahoo, P.R.; Patel, S.; Mishra, B.K. Oxidation of thiourea and substituted thioureas: A review. *J. Sulfur Chem.* **2011**, *32*, 171–197. [CrossRef]
37. Pharmafile. Copper Could Replace Precious Metals in Pharma Production. 2012. Available online: http://www.pharmafile.com/news/175878/copper-could-replace-precious-metals-pharma-production (accessed on 1 November 2021).
38. Tandon, P.; Singh, S. Catalytic applications of copper species in organic transformations: A review. *J. Catal. Catal.* **2014**, *1*, 21–34.
39. Rahman, A.; Uahengo, V.; Daniel, L.S. Green chemistry concept: Applications of catalysis in pharmacuetical industry. *Glob. Drugs Ther.* **2017**, *2*, 1–6.
40. Amphlett, J.; Pepper, S.; Riley, A.; Harwood, L.; Cowell, J.; Whittle, K.; Lee, T.; Ogden, M. Impact of copper(II) on activation product removal from reactor decommissioning effluents in South Korea. *J. Ind. Eng. Chem.* **2019**, *82*, 261–268. [CrossRef]
41. Pepper, S.E.; Whittle, K.R.; Harwood, L.M.; Cowell, J.; Lee, T.S.; Ogden, M.D. Cobalt and nickel uptake by silica-based extractants. *Sep. Sci. Technol.* **2018**, *53*, 1552–1562. [CrossRef]

MDPI
St. Alban-Anlage 66
4052 Basel
Switzerland
Tel. +41 61 683 77 34
Fax +41 61 302 89 18
www.mdpi.com

Eng-Advances in Engineering Editorial Office
E-mail: eng@mdpi.com
www.mdpi.com/journal/eng

www.ingramcontent.com/pod-product-compliance
Lightning Source LLC
LaVergne TN
LVHW070627100526
838202LV00012B/743